Solar Cells and Photocells

Second Edition

by

Rufus P. Turner, Ph. D.

Howard W. Sams & Co., Inc.

4300 WEST 62ND ST. INDIANAPOLIS, INDIANA 46268 USA

International Standard Book Number: 0-672-21711-2
Library of Congress Catalog Card Number: 80-50048

Printed in the United States of America.

Preface

The photocell is a venerable electronic component, its family tree having nineteenth-century roots. Indeed, many historians of electronics regard the selenium cell as the oldest solid-state component, since this device appeared even earlier than the crystal detector which often is accorded the distinction. Applications of the photocell have proliferated in step with the march of electronics; many present-day marvels have been made possible by this relatively simple device which is familiar to the public as the *electric eye*.

This book is primarily a collection of selected practical applications of photocells. Although it is addressed to experimenters, technicians, and science fair participants, other readers no doubt will find it useful. There is a minimum of theory: A brief, but we hope adequate, discussion describes how photocells work and how they are constructed, and this discussion assumes that the reader is already familiar with elementary electronics. Also, Appendix A contains numerous optical and photoelectric terms.

In preparing the material, we have been concerned only with photocells and solar cells, as such. We have left for another time the other familiar photoelectric devices: photodiode, phototransistor, phototube, photodarlington, photo SCR, etc. The circuits have been thoroughly tested and employ only those photocells which at the time of the writing were easily available to hobbyists and experimenters on a one-piece retail basis.

Preparation of the present Second Edition has given us the welcomed opportunity to include additional circuits and to expand the text.

For photographs and/or technical data, I am indebted to Clairex Electronics, Inc.; Edmund Scientific Co.; and International Rectifier Corporation.

RUFUS P. TURNER

Contents

CHAPTER 4

CHAPTER 5

APPENDIX A

APPENDIX B

chapter **1**

Photoelectricity
Simplified

This chapter offers a simple explanation of photoelectric action and describes some of the practical devices that utilize it. Some understanding of this phenomenon is necessary if the reader is to apply photocells and solar cells effectively; however, a knowledge of the advanced physics of photoelectric devices is not demanded. Hence, the presentation here is nonmathematical. Additional background material will be found in the "Glossary of Optical and Photoelectric Terms," which constitutes Appendix A of this book.

1.1 THE PHOTOELECTRIC EFFECT

Some materials (especially the alkali metals—cesium, potassium, and sodium) have the ability to emit electrons in significant quantities when the material is exposed to light. For a particular substance, the number of emitted electrons (called *photoelectrons*) depends upon both the intensity and the wavelength of the light. This action, however, is only one aspect of photoelectricity. It is properly termed *photoemission* and is the basis of the phototube. A second aspect is *photoconductivity*, the phenomenon in which the resistance of certain materials decreases when they are exposed to light. An example of such a material is cadmium sulfide. A third aspect is the *photovoltaic effect,* and this is the phenomenon in which certain materials, properly processed and fabricated into suitable devices, generate a voltage when they are exposed to light. Examples of such materials are selenium and silicon. Whereas photoemission is the basis of the phototube, photoconductivity and photovoltaic action are the basis of photocells, simple light-sensitive devices for the control of electric circuits. This book is concerned only with photocells (of which the solar cell is a particular type) and not at all with phototubes and other photoelectric devices.

All of the principal photoelectric actions were first observed in the 1800's. But practically all of their familiar practical applications have come in the 20th century, and these include television, talking movies, facsimile, alarm devices, object counters, light meters, card and tape readers, object counters, safety devices, light-operated switches and other controls, satellite power supplies, and many others.

1.2 NATURE OF LIGHT

Light is a form of radiant energy; that is, energy which is propagated through space or matter as electromagnetic waves. Light differs from other kinds of electromagnetic radiation— such as radio waves, heat, and X-rays—only in wavelength, or frequency. As shown in Table 1-1, the light spectrum extends from wavelengths of 0.0000001 cm to 0.1 cm, which is the same as 10 to 10,000,000 angstroms. (One angstrom (Å) $= 1 \times 10^{-8}$ cm $= 3.937 \times 10^{-9}$ in.) This range corresponds to frequencies

Table 1-1. Complete Light Spectrum

Wavelength		Frequency
cm	Å	(GHz)
Ultraviolet ⎰0.0000001	10	300,000,000
⎱0.00004	4000	750,000
Visible ⎰		
⎱0.00007	70000	428,570
Infrared ⎰		
⎱0.1	10,000,000	300

Table 1-2. Visible Spectrum

Hue	Wavelength		Frequency (GHz)
	cm	Å	
Violet	0.000041	4100	731,707
Indigo	0.000042	4200	714,286
Blue	0.000047	4700	638,298
Green	0.000052	5200	576,923
Yellow	0.000058	5800	517,241
Orange	0.000060	6000	500,000
Red	0.000065	6500	461,538

from 300,000,000 GHz to 300 GHz, respectively. Immediately below the lower end of the light spectrum lie the ultrahigh-frequency radio waves, and immediately above the upper end lie X-rays.

In a small portion (approximately 0.1 percent) of the nearly 300,000,000-GHz-wide light spectrum, the radiant energy produces the sensation of sight. Table 1-2 shows the approximate wavelengths and frequencies of this visible-light spectrum and identifies the seven hues (colors) which can be seen by an observer. Visible light covers the frequency range 428,600,000 GHz to 750,000,000 GHz (wavelengths of 0.00004 cm to 0.00007 cm, or 4000 Å to 7000Å). The wavelengths and frequencies in Table 1-2 mark the approximate center of each color range. If invisible light (i.e., infrared rays below the visible spectrum and ultraviolet rays above the visible spectrum) is included, the result is the entire light range, which then extends from 0.0000001 cm to 0.1 cm (300 GHz to 300,000,000 GHz).

Although scientists recognized the photoelectric effect, they were long at a loss to explain satisfactorily how light, as waves, managed to dislodge electrons in a light-sensitive material. The Newtonian theory of light (circa 1675) depicted light as composed of material corpuscles, tiny particles projected in a straight line from a luminous body, that struck the eye and produced the sensation of sight. However, this theory was later abandoned in favor of the wave theory. It remained for Albert Einstein to suggest that light is composed of both waves and particles (called *photons*) and that the particles knock electrons out of the atoms of a light-sensitive material. Quantum theory later arrived at the same concept of the dual nature of light (i.e., both wave and particle) and described photons as quanta of luminous energy. For his work, Einstein won the Nobel Prize in 1905.

Light travels at high speed, its velocity in a vacuum being approximately 300,000 kilometers per second (186,000 miles per second) and only slightly lower in air. In other media, light travels somewhat more slowly. For example, light passing through common glass is slowed down to approximately 199,999 km/sec, and through water to approximately 225,000 km/sec. For additional information regarding light, see "Light Source" and related entries in Appendix A.

1.3 PHOTOCELLS, GENERAL

A photocell, also called a *photoelectric cell*, is a solid-state device (usually two-terminal) used to convert luminous energy into electrical energy or to employ luminous energy to control the flow of an electric current. A *photovoltaic cell* (also called a *self-generating photocell*) directly produces a dc voltage that is proportional to incident light. A *photoconductive cell* (sometimes called a *photoresistive cell*) changes its internal resistance inversely with the intensity of incident light and accordingly can vary the strength of an electric current. In this cell, the ratio of dark resistance to light resistance can be extremely high.

Photocells employing selenium or silicon as the light-sensitive material can function as either photovoltaic or photoconductive devices; however, the silicon type is used almost exclusively in the photovoltaic mode, owing to its higher output voltage for a given illumination. Historically, an early photovoltaic cell employing a film of copper oxide on a copper plate was supplanted by the selenium cell as we now know it. Light-sensitive materials employed in most photoconductive cells include cadmium selenide, cadmium sulfide, indium antimonide, and lead sulfide. Photoconductive cells and photovoltaic cells are made in a wide variety of sizes and shapes, and conform to a wide range of electrical specifications.

The spectral response (see Glossary, Appendix A) of the selenium photocell peaks at approximately 5550 angstroms and that of the silicon cell peaks at approximately 8000 angstroms;

both of these devices are superior to the human eye in their response. Selenium and silicon cells both are employed in color meters and color matchers and in chemical analyzers based upon substance color. The cadmium-selenide cell shows peak response at approximately 6900 angstroms, and the cadmium-sulfide cell at approximately 5500 angstroms.

Largely because of their (usually) flat-plate type of construction, many photocells exhibit significant amounts of shunting capacitance (typically 0.35 μF/sq in for silicon units, 0.17 μF/sq in for selenium units), and this limits their high-frequency response unless the cell is quite small. For applications in which the effects of this capacitance are detrimental —such as high-frequency sound recording and reproduction and very-high-speed counting and switching—photocells may prove unsuitable and require replacement with photodiodes or phototransistors.

The photoconductive cell is described in more detail in Section 1.4, and the photovoltaic cell in Section 1.5. Section 1.7 describes the range of electrical characteristics of photocells.

1.4 PHOTOCONDUCTIVE CELL

The photoconductive cell (sometimes called a *photoresistive cell*) is one in which the light-sensitive material lowers its resistance (increases its conductance) in proportion to an increase in incident light. When such a cell is operated in series with a load and a voltage source (such as a battery or power supply), it acts as a light-variable resistor, the current flowing in the circuit increasing with illumination, and vice versa. There are many variations of this simple, basic circuit.

Commercial photoconductive cells use either cadmium selenide, cadmium sulfide, or lead sulfide as the light-sensitive material, with cadmium sulfide being the most prominent. As mentioned in Section 1.3, selenium photocells and silicon photocells can be used as photoconductive devices, but they rarely are, since they are valued more for their photovoltaic properties.

Photoconductive cells do not employ junctions, but utilize the bulk resistance effects of a suitable light-sensitive material. For this purpose, two electrodes are attached to, or imbedded

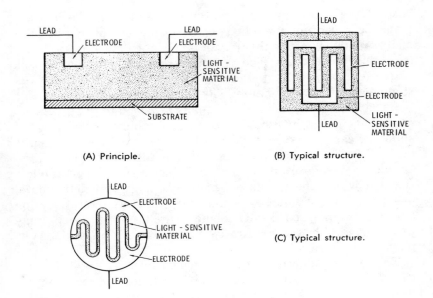

(A) Principle.

(B) Typical structure.

(C) Typical structure.

Fig. 1-1. Photoconductive-cell structure.

into, the material (see simplified version in Fig. 1-1A). However, in most such cells, more-complicated electrodes are employed to provide greater contact length while still leaving room for ample light entry to the sensitive material between the electrodes (see Figs. 1-1B and 1-1C). Fig. 1-2 shows some typical photoconductive cells.

1.5 PHOTOVOLTAIC CELL

The photovoltaic cell (also called a *self-generating cell*) is one that generates an output voltage in proportion to the intensity of incident light. This device, therefore, is a direct converter of luminous energy into electrical energy. Modern commercial photovoltaic cells are of either the selenium or the silicon type. All photovoltaic cells are junction-type devices.

Selenium Cell—In this photovoltaic cell, the light-sensitive material is specially processed selenium. The cross section of a selenium cell resembles that of a selenium rectifier plate, the actual construction varying somewhat, however, with different

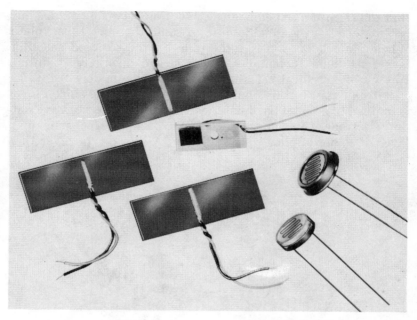

Fig. 1-2. Typical photoconductive cells.

manufacturers. In general, the selenium is applied to a metal base plate which becomes the positive terminal of the device; the selenium film itself is the negative electrode, and an ohmic contact is usually made to it by means of a sprayed-on metal strip applied near the edge of the film. Fig. 1-3A gives a simplified picture of this arrangement. Note from this construction

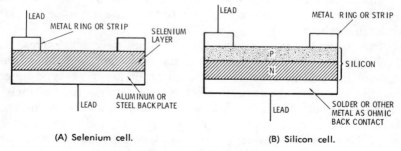

(A) Selenium cell.   (B) Silicon cell.

Fig. 1-3. Basic structure of photovoltaic cells.

that the selenium cell embodies a junction between two dissimilar materials—the selenium layer and the metal backplate. When the cell is in use, light falls on the exposed selenium layer. At 2000 footcandles, the average open-circuit output voltage of a typical selenium cell is approximately 0.45 volt.

Silicon Cell—In this photovoltaic cell, an n-type silicon layer is applied to a metal backplate which becomes the negative output electrode (in some models, this backplate consists of a layer of solder applied to the silicon). A thin p-type layer then is formed on, or diffused into, the exposed face of the n-type

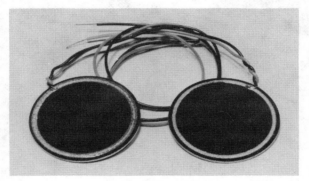

Courtesy International Rectifier Corp.
Fig. 1-4. Typical photovoltaic cells (unmounted).

layer. Finally, for ohmic contact, a sprayed-on (or plated) strip or ring is applied to the p-type layer. This sprayed-on strip or ring becomes the positive output electrode. Fig. 1-3B gives a simplified picture of this arrangement. In some models of silicon cell, the n-type layer is on top, the p-type layer on the bottom, and the output-voltage polarity is the opposite of that described above.

The n- and p-type layers form a relatively large-area pn junction across which there is a natural electric field. Light energy impinging upon the exposed outer layer (p-type layer in this instance) generates electron-hole pairs and minority carriers (electrons in the p-type layer and holes in the n-type layer), and electrons are swept from p to n, and holes from n to p. This action produces the output voltage of the cell, the p-type layer being positive and the n-type layer negative. At

2000 footcandles, the average open-circuit output voltage of a typical silicon cell is approximately 0.3 to 0.6 volt.

Fig. 1-4 shows typical, unmounted photovoltaic cells which may be of either the selenium or the silicon type.

1.6 SOLAR CELL

Generally speaking, a solar cell is a heavy-duty photovoltaic cell; that is, any self-generating cell that can produce usefully high voltage and current when exposed to sunlight. A typical solar cell, however, is a silicon cell; this type of photovoltaic cell delivers the highest output for a given light intensity.

A *solar battery* is a dc power source made up of several solar cells connected in series or parallel, or both, to deliver useful amounts of power when illuminated by sunlight. Such heavy-duty photoelectric batteries are used in space satellites, control devices, emergency telephone power supplies, portable radios, and other places. Fig. 1-5 shows a solar battery being used to

Courtesy Edmund Scientific Co.

Fig. 1-5. Solar-cell panel charging a 12-volt battery.

trickle charge a 12-volt storage battery. In this case, thirty ½-volt silicon photovoltaic cells are connected in series for voltage multiplication; a series diode prevents backflow of current from the battery through the cells during darkness. In bright sunlight, this unit delivers 0.1 A at 12 V and has a capacity of 30 watt-hours per week.

1.7 RANGE OF CHARACTERISTICS

Below are listed the principal electrical characteristics of photoconductive cells and photovoltaic cells and the range of these characteristics in commercially available cells. A number of variables influence these typical characteristics which depend finally upon make, model, and geometry of the cell and upon the composition of the photosensitive material. The listing, therefore, is intended only to show the obtainable values in each characteristic, and the entries accordingly are not necessarily parallel. For example, a maximum current of 36 mA and a maximum output voltage of 12 V are shown for a conventional silicon cell, but this does not necessarily mean that a 12-V solar cell delivers 36 mA.

Photoconductive Cell

Maximum applied voltage—The voltage drop across the cell: for cadmium sulfide, 20 V to 300 V; for cadmium selenide, 100 V to 300 V.

Maximum power dissipation—The product EI, where E is the voltage drop (volts) across the cell, and I is the cell current (amperes). For cadmium selenide and cadmium sulfide at 25°C, the maximum power dissipation is 50 mW to 2 W. Derating curves are available, and heat-sink operation is recommended in some instances.

Dark resistance—The internal resistance of the cell when the latter is completely shielded from light: for cadmium sulfide, 1.6K to 1000 megohms; for cadmium selenide, 120K to 3000 megohms.

Light resistance—The internal resistance of the cell when the latter is exposed to a specified illumination: for cadmium sulfide, 0.11 ohm at 100 footcandles (fc), 1.5K to 700K at 2 fc; for cadmium selenide, 1.5K to 133K at 2 fc.

Peak spectral response—For cadmium sulfide, 5500 to 6200 angstroms. For cadmium selenide, 6900 to 7350 angstroms.

Photovoltaic Cell

Output current—Varies with active area, illumination, and load resistance. The typical 100-fc short-circuit value for selenium is 12 μA to 770 μA; for silicon, 5 mA to 36 mA; for silicon solar-power assemblies, 18 mA to 500 mA.

Output voltage—Open-circuit. Directly proportional to active area and illumination. Typical 100-fc value for selenium is 0.2 V to 0.45 V; for silicon, 0.3 V to 1.5 V; for silicon solar-power assemblies, 0.4 to 12 V.

Power output—Varies with active area, illumination, and load resistance. For a given illumination and active area, the power output is maximum when the load resistance equals the internal resistance of the cell. When the cell output is pure dc, the power (in watts) is the simple product EI, where E is the output voltage (in volts) and I the output current (in amperes).

Internal resistance—For a given illumination, the dc resistance

$$R = E/I$$

where,
E is the output voltage (volts) developed across the cell,
I is the output current (amperes),
R is in ohms.

The resistance is lower in a silicon cell than in an equivalent selenium cell.

Peak spectral response—For selenium, 5500 angstroms. For silicon, 8000 angstroms.

1.8 BASIC CIRCUITS

There are only three basic circuits for photocells, and these are shown in Fig. 1-6; all others are elaborations of these. In each of the basic circuits, the load device is shown as a simple resistor, R_L, but can be any device—such as a meter, relay, counter, lamp, bell, etc.—which is capable of using the output of the photocell (PC).

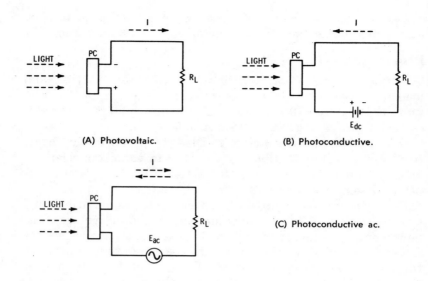

(A) Photovoltaic.

(B) Photoconductive.

(C) Photoconductive ac.

Fig. 1-6. Basic photocell circuits.

In Fig. 1-6A, the light-generated output of the photovoltaic cell is applied directly to the load. If the light is steady, the output is dc; if the light is chopped or is modulated in intensity, the output is pulsating dc or is ac superimposed upon dc. In Fig. 1-6B, the photoconductive cell is biased with a dc voltage (battery or power supply), E_{dc}. The light-controlled resistance of the cell varies the current (I) flowing from E_{dc} through load device R_L. If the light is steady, the current is pure dc; if the light is chopped or is modulated in intensity, the current is composite (i.e., consists of ac superimposed upon dc). In Fig. 1-6C, the photoconductive cell is biased with an ac voltage E_{ac}. Again, the light-controlled resistance of the cell varies current I through the load device, but this time the current is ac even if the light is steady.

1.9 HISTORICAL NOTE

In 1887, Heinrich Hertz, who is best known for his discovery of radio waves, observed that his receiving spark gap would fire more readily when the gap was illuminated with ultraviolet light. He did not understand why this should happen. In 1888, however, William Hallwachs explained the phenomenon, announcing that the ultraviolet light caused negative electricity to be emitted from the metal balls of the gap and that this electricity facilitated the sparking. He proved his point by connecting a polished zinc plate to an electroscope, charging the combination, and exposing the surface of the plate to ultraviolet light. When the plate and electroscope were positively charged, the ultraviolet light had no effect; but when they were negatively charged, the ultraviolet light discharged the combination (thus, negative electricity passed from the plate into the surrounding air). This is the first recorded observation and explanation of the photoelectric effect, which is sometimes called the *Hallwachs effect*. In 1899, Lenard and Thomson identified Hallwachs' negative electricity as a stream of electrons.

In 1873, the photoconductivity of selenium was discovered by Willoughby Smith and by May. The photovoltaic effect in selenium was separately discovered by Adams and Day in 1876 and by Fritts in 1884. Becquerel's later discovery (1899) concerned electrolytic devices. He observed that a voltage appears between two similar electrodes immersed in an electrolyte, such as lead nitrate, when one of the electrodes is exposed to light.

A copper-oxide photovoltaic cell, known by the trade name *Photox cell* and developed by Bruno Lange, was introduced during World War I and marketed by Westinghouse. The predecessor of the present-day selenium cells, *Photronic cell*, appeared during the 1930s and was marketed by Weston.

As noted in Section 1.2, Albert Einstein received the Nobel Prize in 1905 for his explanation of the photoelectric effect. He showed, among other things, that the energy of the fastest photoelectrons is directly proportional to the frequency of the incident light:

$$\tfrac{1}{2} mv_m^2 = hv - \phi$$

where,

  h is Planck's constant (6.63×10^{-27} erg-second),
  m is the mass of the electron,
  v is the frequency of incident light,
  v_m is the velocity of the fastest electron,
  ϕ is the energy in ergs required by an electron to escape from the surface of the photoelectric emitter.

It was not until 1958 that Chapin, Fuller, and Pearson—working at Bell Telephone Laboratories—discovered the high efficiency of the silicon photovoltaic cell, thus paving the way to modern solar cells.

1.10 FUTURE OF PHOTOELECTRICITY

The substantial progress already made in applications of photocells and solar cells in both casual and sophisticated electronics points to expanding use of these devices in commercial, scientific, industrial, military, and household areas. Some of the fields in which it is easy to envision further use of photoelectricity are automatic control; crime detection and prevention; identification, sorting, and grading; counting; communications; safety measures; highway traffic management; pollution

Courtesy International Rectifier Corp.

Fig. 1-7. Experimental sun-powered automobile.

measurement and control; medical technology; and sports and amusements.

One of the most engrossing prospects is that of harnessing sunlight for the production of electrical energy. Experts are fairly confident that at least part of the electricity required in the future will be supplied by rooftop-mounted photovoltaic panels. It is estimated that the earth receives approximately 126 trillion horsepower (94 trillion kilowatts) from the sun each second; and if this enormous and unlimited energy could be more fully utilized, our dependence upon fossil fuels might be dramatically reduced. In this direction, Fig. 1-7 shows an early experimental electric automobile which is sun powered. Here, the large panel seen atop the 1912 Baker car contains 10,640 silicon photovoltaic cells. Its output converts sunlight into enough electricity to keep charged the storage batteries that run the car.

Since the first edition of this book appeared, research and development have been stepped up considerably. This acceleration has been occasioned in great part by shortages of crude oil. Government and industry both have feasibility studies under way, and in 1979, President Carter called for 20 percent of US energy needs to be supplied by solar means by the year 2000. Several energy systems have been suggested to reduce or even to end our dependence on oil and other fossil fuels. Under serious consideration are nuclear, solar-steam, solar-hydrogen, geothermal, hydroelectric, wind, and tidal systems. Of these, various solar systems are attractive, since the sun is a perpetual source of energy. And of the solar systems, the photovoltaic method seems most desirable if it can be made economically feasible, since it converts solar energy *directly* into electricity, is pollution free, and generates no harmful wastes.

Many practical problems remain to be solved, however, before photovoltaic systems can become cost competitive with coal-fueled, oil-fueled, and nuclear systems for large blocks of power. At the time of this writing, authorities assert that solar-cell–produced energy costs ten times as much as that produced from coal. At present, photoelectric power also demands a high capital investment.

In the direction of improvement of the silicon solar cell,

additives—such as chlorophyl—are being tested and promise to increase the efficiency of this device. More efficient storage batteries will be needed, and some improvement must be planned for dc-to-ac converters for use in conjunction with solar cells. One expectation is that by 1986 improved photo-electric energy systems will supply electricity at the rate of 70 cents per peak watt, in terms of 1980 dollars.

It has been estimated that 1000 square feet of solar cells will be required to power the average house. But, because direct solar power is unavailable at night and on cloudy days, it might require a conventional-system backup. There are two approaches to photovoltaic power: central-plant and individual installations. It has even been suggested that during the day when occupants are away from home and do not need the service, a residential installation might supply power, for which the local utility would pay, into the grid; at night, the customer would pay for power from the local utility.

chapter **2**

Light Meters

The light meter—known by this and various other names, depending upon its end use—is perhaps the simplest and most obvious application of the photocell. Both photovoltaic and photoconductive cells are used in light meters.

Light meters are used principally for measuring light intensity in illumination surveys, for testing of lamps and other light sources, and in photography. In various forms, however, they are used also in analyzing and matching colors, checking the turbidity of solutions, checking the density of smoke or fog, making reflection and glare tests, and many kindred applications.

This chapter describes thirteen practical circuits for light meters, including selenium, silicon, and cadmium-sulfide types. As for electrical characteristics, representative photocells are shown; and while the circuits may be used as shown, they may also be readily adapted for use with other photocells of the same variety in the reader's possession.

2.1 LIGHT-METER CALIBRATION

The calibration of a light meter will pose a problem to the average hobbyist or experimenter, since few accurate light sources are available for his use. Light standards usually are found only in photometric laboratories. The serious builder of a light meter must therefore either arrange to have his instrument calibrated by a laboratory offering this service, must employ whatever approximate light source or already calibrated meter (such as a suitable photographic exposure meter) is available to him, or must be content to use his instrument, at least for the time being, only for comparative measurements. The light sensitivities given for circuits in this chapter apply for the particular circuit the author assembled and tested, and should be reasonably close for the reader's version of the circuit.

Some photographic exposure meters have a meter scale reading directly in footcandles (fc), and such an instrument will be useful, as far as it goes, for the calibration of a homemade light meter. However, a full-scale deflection of 100 fc or less is usually the limit of such meters. When an exposure meter is used, its photocell and the one in the homemade instrument must be placed as close together, side by side, as practicable, during the calibration, so that both instruments experience the same illumination. For this purpose, an adjustable light source can be made with a 150-watt incandescent lamp operated through a variable transformer. Under favorable circumstances, a few points may be checked in each range of a homemade meter, by using this method.

Some idea of the approximate performance of a circuit may be gained by consulting the performance curves supplied by the manufacturer of the photocell used. These curves show short-circuit current and/or open-circuit voltage versus illumi-

Table 2-1. Pilot-Lamp Light Values

Type	Volts	Footcandles
6ESB	6	250
10ESB	10	450
12ESB	12	650
24ESB	24	1100
28ESB	28	1285
48ESB	48	2000
93	12	15
1133	6	32
1156	12	32

nation. However, various circuit factors must be taken into account in interpreting these data for the reader's circuit. These factors include the resistance in series with the cell, and the internal resistance of the indicating meter.

It is helpful to know that the approximate value of light obtained with various electric lamps may be found in lamp manufacturers' literature. Thus, a 60-watt tungsten-filament lamp delivers 50 footcandles at a distance of 1 foot. Table 2-1 shows the approximate footcandles obtained with various pilot lamps.

2.2 SELENIUM LIGHT METER

Fig. 2-1 shows the circuit of the least-expensive light meter; this circuit employs a single B3M-C selenium photocell (PC) to drive directly a 0 to 50 dc microammeter (M).

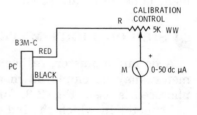

Fig. 2-1. Selenium light meter.

Because the dc output of the selenium cell is relatively low for a given light intensity, the sensitive meter is required, and an illumination of approximately 150 footcandles is needed for full-scale deflection. Calibration of the instrument is simple: With the cell illuminated with light of known intensity and shielded from daylight or room light, adjust the 5000-ohm wirewound rheostat (R) for exact full-scale deflection. The rheostat then needs no readjustment unless its setting is accidentally disturbed or the instrument is being recalibrated.

2.3 SILICON LIGHT METER WITH MICROAMMETER

Use of a low-output silicon photocell allows a more rugged microammeter to be used than the one in the circuit described in Section 2.2. Fig. 2-2 shows the circuit of a light meter employing a 0 to 100 dc microammeter with an S1M-C silicon

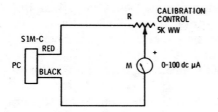

Fig. 2-2. Silicon light meter.

photocell (PC). An illumination of approximately 60 footcandles gives full-scale deflection.

The instrument is easily calibrated: With the cell illuminated with light of known intensity and shielded from daylight or room light, adjust the 5000-ohm wirewound rheostat (R) for exact full-scale deflection. The rheostat then needs no readjustment unless its setting is accidentally disturbed or the instrument is being recalibrated.

2.4 SILICON LIGHT METER WITH MILLIAMMETER

A dc milliammeter, rather than the more delicate microammeter, may be employed in a light meter if a higher-output photocell is used. Fig. 2-3, for example, shows the circuit of a light meter in which the indicator is a 0 to 1 dc milliammeter (M) driven by an S4M-C silicon photocell (PC). An illumination of only 70 footcandles gives full-scale deflection of the meter.

Calibration of the circuit is easy and simple: With photocell PC illuminated with light of known intensity and shielded from daylight or room light, adjust the 5000-ohm wirewound rheostat (R) for exact full-scale deflection. The rheostat then needs no readjustment unless its setting is accidentally disturbed or the instrument is being recalibrated.

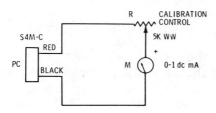

Fig. 2-3. Silicon light meter with milliammeter.

When the comparative ruggedness of the milliammeter is not needed, the sensitivity of this light-meter circuit may be increased by using a dc microammeter instead. With a 0 to 100 dc microammeter, for example, full-scale deflection occurs with an illumination of approximately 40 footcandles at the input.

2.5 HIGH-CURRENT-OUTPUT SILICON LIGHT METER

When only a hefty milliammeter is available, a high output current will be required from a photocell. The cell current may be boosted by means of a simple transistor-type dc amplifier. An amplifier-type light meter employing a 0 to 50 dc milliammeter is shown in Fig. 2-4.

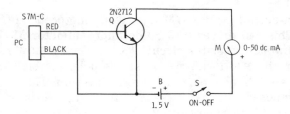

Fig. 2-4. High-current-output silicon light meter.

In this arrangement, the output of an S7M-C silicon photocell (PC) is amplified by an inexpensive 2N2712 silicon transistor (Q). Operating power for the transistor is supplied by a single, 1.5-V, size-D cell (B). The static collector current (less than 1 μA with zero-light input) is so low that no zero-set potentiometer is needed.

The maximum rated output of the S7M-C cell is 8 mA. With the simple amplifier, however, an illumination of approximately 1000 footcandles drives the milliammeter to its full-scale deflection of 50 milliamperes.

This circuit is useful not only for its accommodation of a high-range milliammeter, but also for its possible interface with a high-current (low-impedance) electromechanical recorder. The recorder input terminals may be connected directly in place of meter M.

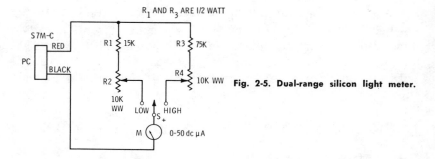

Fig. 2-5. Dual-range silicon light meter.

2.6 DUAL-RANGE SILICON LIGHT METER

For convenience and versatility, the circuit shown in Fig. 2-5 provides two ranges. The LOW range gives a coverage of approximately 0 to 100 fc and the HIGH range approximately 0 to 1000 fc.

A high-output, type S7M-C silicon photocell (PC) is employed. This cell drives 0 to 50 dc microammeter M through a selected limiting-resistor circuit (R_1-R_2 for the LOW range, and R_3-R_4 for the HIGH range). A single-pole, double-throw non-shorting switch (S) permits range selection. Each resistance leg is composed of a fixed (safety, limiting) resistor (R_1, R_3) and a wirewound rheostat (R_2, R_4), the latter serving as the calibration control for its range.

A two-step calibration is required:

1. With switch S set to LOW, illuminate photocell PC with approximately 100 footcandles and adjust rheostat R_2 for exact full-scale deflection of meter M.
2. With switch S set to HIGH, illuminate photocell PC with approximately 1000 footcandles and adjust rheostat R_4 for exact full-scale deflection of meter M. After these calibration adjustments, the two rheostats will require no further adjustment unless their settings are accidentally disturbed or the instrument is being recalibrated.

2.7 TRANSISTORIZED SILICON LIGHT METER

The silicon solar cell is basically a low-voltage, high-current device. In some applications, especially where high impedances are involved, more convenient action might be obtained if the dc output voltage of the cell could be boosted and the cell looked like a high-impedance source. For example, the photocell might drive a dc voltmeter through an amplifier.

Fig. 2-6 shows the circuit of a simple dc amplifier based upon a single U183 field-effect transistor, Q. The maximum no-load output (0.4 V at approximately 1000 fc) of the S1M-C photocell is amplified to 1.6 V by this arrangement. A particular advantage of a circuit such as this is the high-resistance load that the FET input offers to the photocell. On rapidly changing light signals and those modulated at a high frequency, this input avoids the high damping caused by the low resistance into which such cells usually operate.

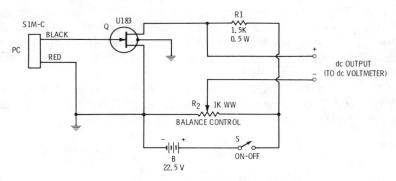

Fig. 2-6. Transistorized silicon light meter.

The output half of the circuit is a resistance bridge whose four arms are R_1, the internal drain-to-source resistance of the FET, and the two "halves" of potentiometer R_2. With the photocell completely darkened, R_2 is set to balance out the no-signal static voltage that appears at the dc output terminals (an electronic dc voltmeter temporarily connected to these terminals will serve as the balance indicator). At the balance point, the meter reads zero, and the circuit should then remain balanced indefinitely.

Subsequent illumination of the cell will unbalance the bridge because the dc output of the cell then changes the drain-to-source resistance of the FET. This causes 1.6 volts to appear at the dc output terminals (this voltage level will maintain itself across an external load of 20,000 ohms). The total maximum-signal current drawn from the 22.5-V source (battery B) is approximately 24 mA.

2.8 TRANSISTORIZED SELENIUM LIGHT METER

The circuit shown in Fig. 2-7 boosts the output of a small, B3M-C selenium photocell (PC) sufficiently to allow use of a 0 to 1 dc milliammeter (M). Compare this circuit with the simpler one described in Section 2.2 and illustrated by Fig. 2-1.

In this sensitized arrangement, a 2N2608 field-effect transistor (Q) is employed as a dc voltage amplifier to boost the 0.85-V maximum output of the selenium cell. A four-arm bridge circuit (R_1, R_2, the internal drain-to-source resistance of the FET, and 5000-ohm wirewound rheostat R_3) is used to balance out the initial flow of FET drain current through milliammeter M when the photocell is darkened. This entire circuit acts as a FET-type electronic dc voltmeter measuring the output of the photocell.

This arrangement increases the photocell sensitivity from 2 to 3 times, depending upon individual FET and selenium cell characteristics.

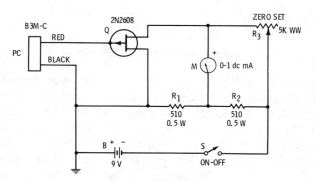

Fig. 2-7. Transistorized selenium light meter.

2.9 BASIC CADMIUM-SULFIDE LIGHT METER

A light meter employing a photoconductive cell has the advantage of good sensitivity—which means that small light values can be measured with a comparatively rugged indicating meter. A slight disadvantage is the requirement of a self-contained battery; however, this is no great inconvenience.

Fig. 2-8 shows the circuit of the simplest light meter employing a cadmium-sulfide photocell. Here, the cell is a type CS120-C operated from a 1.5-V, size-D cell (B), and the indicating meter is a 0 to 1 dc milliammeter, M. The dark current of the photocell is too low to be seen on the meter scale. Full-scale deflection is obtained with an illumination of approximately 10 footcandles. Changing rheostat R to 1500 ohms

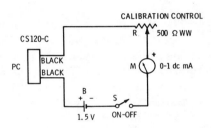

Fig. 2-8. Basic cadmium-sulfide light meter.

makes full-scale deflection 1000 fc; and changing B to 9 V, with R at 500 ohms maximum, makes full-scale deflection 1 fc. A number of footcandle ranges can be obtained by switching both the battery voltage and the rheostat resistance.

Calibration is simple and easy: With photocell PC illuminated with approximately 10 footcandles and protected from daylight or room light, adjust CALIBRATION CONTROL rheostat R for exact full-scale deflection of milliammeter M. After this, the rheostat will need no further adjustment unless its setting is accidentally disturbed or the instrument is being recalibrated.

2.10 HIGH-OUTPUT BRIDGE-TYPE LIGHT METER

Fig. 2-9 shows the circuit of a cadmium-sulfide light meter using a conventional 20,000-ohms-per-volt voltmeter as the indicator. With this circuit, a light input of 20 footcandles to the CS120-C cadmium-sulfide photocell (PC) produces a deflection of 7 volts. The voltmeter may be switched to lower ranges for greater light sensitivity.

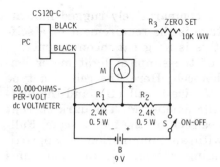

Fig. 2-9. High dc output bridge-type
light meter.

The circuit consists of a four-arm bridge (R_1, R_2, the internal resistance of photocell PC, and wirewound rheostat R_3). If the photocell resistance is denoted by R_c, the rheostat setting required to balance the bridge is $R_3 = (R_c R_2) / R_1$.

Initial adjustment of the instrument is straightforward and is the same as zero-setting an electronic voltmeter: With the photocell darkened, balance the bridge by adjusting rheostat R_3 for zero reading of voltmeter M. The circuit is then ready for use: Illuminate the cell to obtain voltmeter deflection proportional to the intensity of the incident light. To change light ranges, simply change the voltmeter range. If the bridge is balanced with the meter switched to its lowest voltage range, it will automatically be balanced (zeroed) on the higher ranges.

2.11 BASIC AC LIGHT METER

It sometimes is desirable for a light-meter circuit to actuate an ac meter. One reason for this is the availability of high-input-impedance electronic ac voltmeter/millivoltmeters with full-scale deflections down to 1 mV. Such a meter enables measurement of very-low light levels.

Fig. 2-10 shows a suitable 60-Hz ac circuit employing a CS120-C cadmium-sulfide photocell (PC) biased by 6.3 volts from a small filament-type transformer, T. When the cell is darkened and switch S is closed, virtually no current flows from the secondary winding of the transformer and through output resistor R; the photocell thus acts as an open switch. Upon illumination, however, the internal resistance of the cell

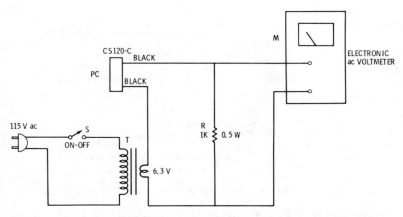

Fig. 2-10. Basic ac light meter.

drops in proportion to the intensity of the light, and a proportionate alternating current flows through resistor R, developing across this resistor a voltage drop which is proportional to the light intensity and which is indicated by the electronic ac voltmeter (M). With illumination of approximately 70 footcandles, the deflection of the meter is 5 V rms. The meter is easily switched to lower ranges for greater light sensitivity, and vice versa.

The arrangement shown in Fig. 2-10 operates at the power-line frequency—usually 60 Hz. However, it can be modified for operation at other audio frequencies by substituting an audio-grade transformer with a suitable turns ratio for the filament transformer shown in the diagram, and by feeding in a constant-voltage signal from an audio oscillator.

2.12 CADMIUM-SULFIDE AC LIGHT METER WITH RATIOMETER-TYPE BRIDGE

A variation of the cadmium-sulfide ac light meter is shown in Fig. 2-11. This circuit, like the one described in Section 2.11, employs an electronic ac voltmeter/millivoltmeter (M) as the indicator. This circuit embodies a bridge composed of wire-wound rheostat R, the internal resistance of the CS120-C photocell (PC), and the two halves of the secondary winding of the small, 6.3-V filament-type transformer, T.

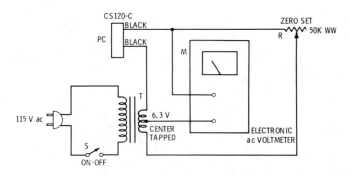

Fig. 2-11. Ratiometer-bridge-type ac light meter.

With switch S closed and the cell darkened, the bridge is balanced by adjusting rheostat R for zero deflection of meter M. If this balancing operation is performed with the meter switched to its lowest-voltage scale, the zero setting will be maintained automatically on the higher ranges. When the photocell later is illuminated, a light input of approximately 20 footcandles will result in a 2.2-V rms deflection. The meter is easily switched to lower ranges for greater light sensitivity, and vice versa.

The arrangement shown in Fig. 2-11 operates at the power-line frequency—usually 60 Hz. However, it can be modified for operation at other audio frequencies by substituting an audio-grade transformer with a suitable turns ratio and a center-tapped secondary for the filament transformer shown in the diagram, and by feeding in a constant-voltage signal from an audio oscillator.

2.13 CADMIUM-SULFIDE LIGHT-TO-AC TRANSDUCER FOR PHOTOMETRY

In some photometric measurements, a self-contained indicating meter, such as that shown in most of the preceding circuits, is not desired. Instead, the output of the light-meter circuit must be applied to a potentiometer, recorder, oscilloscope, comparator, data processor, or other external device. Fig. 2-12 shows the circuit of an ac-type cadmium-sulfide circuit for this purpose.

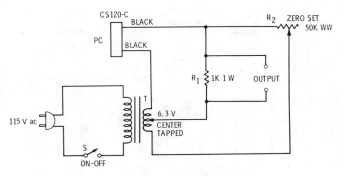

Fig. 2-12. Light-to-ac transducer circuit.

This circuit, like the one described in Section 2.12, embodies a ratiometer-type bridge. With switch S closed and the CS120-C photocell (PC) darkened, the bridge is balanced by adjusting rheostat R_2 for zero deflection of the output device connected to the OUTPUT terminals. When the cell later is illuminated, a light input of approximately 30 footcandles results in a no-load output voltage of 1.5-V rms across resistor R_1. Stronger illumination will give higher output voltage, and vice versa.

The arrangement shown in Fig. 2-12 operates at the power-line frequency—usually 60 Hz. However, it can be modified for operation at other audio frequencies by substituting an audio-grade transformer with suitable turns ratio and center-tapped secondary for the filament transformer shown in the diagram, and by feeding in a constant-voltage signal from an audio oscillator.

2.14 HIGH-SENSITIVITY, IC-TYPE LIGHT METER

Fig. 2-13 shows the circuit of a light meter for measuring very low levels of illumination. The 741 operational-amplifier IC serves as a gain-of-100 dc amplifier to boost the output of the S1M-C self-generating photocell (PC) before deflection of the 0-1 dc milliammeter, M.

In many ICs, the offset voltage is so low that no darkened-cell zero set of the meter is necessary before taking light readings. With others having offset high enough to deflect the meter with the cell darkened, R_2 may be added.

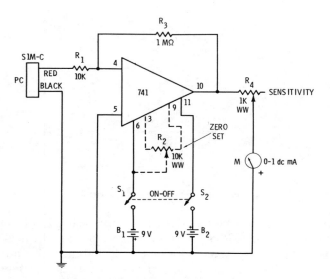

Fig. 2-13. High-sensitivity, IC-type light meter.

To check the circuit: (1) With SENSITIVITY rheostat R_4 set arbitrarily to center range, illuminate photocell PC, noting that meter M deflects upscale. (2) If a light source of accurately known intensity is available, use it to illuminate the cell steadily, and set rheostat R_4 for exact full-scale deflection of the meter. (3) Vary the intensity of the light while keeping constant the distance between source and cell, and note corresponding deflections of the meter.

For separate ranges of the instrument, experimentally determined values of either R_3 or R_4 may be switched into the circuit. Increasing R_3 increases the sensitivity; increasing R_4 decreases the sensitivity.

chapter **3**

Relays and Control Circuits

Light-operated relays find many applications, both alone and as part of other equipment. Some, like the one shown in Fig. 3-1, consist simply of a photovoltaic cell driving a sensitive dc relay directly; others (for example, Fig. 3-2) employ a photoconductive cell and require a dc supply; still others, as in Fig. 3-8, employ an ac supply with a photoconductive cell; and many more (as in Fig. 3-3 and other circuits in this chapter) use a transistor to boost the output of the photocell either to provide a more sensitive circuit or to operate a higher-current relay. A few of the devices in which these basic relays are used are automatic light switches, burglar alarms, fire alarms, game machines, object counters, remote controls, safety devices, and smoke controls.

Most of the circuits are simple. Where a multistage amplifier is required, an integrated circuit is shown (as in Figs. 3-7, 3-9, 3-10, and 3-11) to save wiring labor. Although a battery is shown in most of the circuits, a well-filtered power-line-operated dc power supply can be used instead. For names and locations of manufacturers of the photocells and other special components shown in this chapter, see Appendix B.

3.1 DIRECT-OPERATING SILICON-CELL RELAY

The output of a photovoltaic cell can directly drive a sufficiently sensitive dc relay—no battery or power supply is needed. This is the ultimate in simplicity.

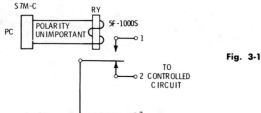

Fig. 3-1. Direct-operating silicon-cell relay.

Fig. 3-1 shows such a relay circuit. Here, the S7M-C silicon photocell supplies enough direct current to close the type 5F-1000S 1-mA, 1000-ohm relay (RY). The polarity of the cell is unimportant; the cell leads may be reversed without malfunction. Light from a 75-watt incandescent lamp at a distance of 1 foot (or equivalent illumination) will close the relay.

The relay contacts will handle 0.25 ampere. This current rating will allow some devices to be operated directly from the relay. In other instances, the light-operated relay may be used to operate an external heavier-duty relay. Operation may be obtained with the light beam on or off: Use relay contacts 1 and 3 for beam-on operation; use 2 and 3 for beam-off operation.

3.2 BASIC PHOTOCONDUCTIVE-CELL RELAY

A photoconductive cell requires a battery or dc power supply. But, for this small inconvenience, it provides good light sensitivity, and sometimes allows a higher-current relay to be used than otherwise would be possible with a small photovoltaic cell. Fig. 3-2 shows the basic circuit for photoconductive cells, and in this instance uses a CS120-C cadmium-sulfide cell (PC).

The cell is biased by a 9-V battery (B) and acts as a light-variable resistor, controlling current through the type 5F-1000S 1-mA, 1000-ohm relay (RY). The polarity of the bat-

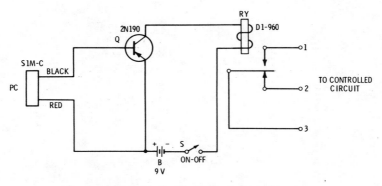

Fig. 3-2. Basic cadmium-sulfide-cell relay.

tery, photocell, and relay is unimportant. An illumination of approximately 5 footcandles will close the relay.

The relay contacts will handle 0.25 ampere. This current rating will allow some devices to be operated directly from the relay. In other instances, the light-operated relay may be used to operate an external heavier-duty relay. Operation may be obtained with the light beam on or off: Use relay contacts 1 and 3 for beam-on operation; use 2 and 3 for beam-off operation.

3.3 TRANSISTORIZED SILICON-CELL RELAY (HIGHER-CURRENT RELAY)

The comparatively low output of a small silicon photocell may be amplified, by means of a simple transistorized dc amplifier, to operate a less-expensive, higher-current relay than the cell alone could operate. Fig. 3-3 shows a transistor circuit of this kind.

Fig. 3-3. Transistorized silicon-cell relay (10-mA relay).

In this arrangement, a single 2N190 transistor (Q) in a common-emitter circuit amplifies the output of the S1M-C photocell (PC) to operate a D1-960 10-mA, 100-ohm relay (RY). The photocell must be polarized so that its negative (black) output lead is connected to the base of the transistor. A 9-V battery (B) supplies the required dc power. Illumination of approximately 70 footcandles closes the relay.

The relay contacts will handle 0.5 ampere at 3-V dc or 12-V ac. This current rating will allow some devices to be operated directly from the relay. In other instances, the light-operated relay may be used to operate an external heavier-duty relay. Operation may be obtained with the light beam on or off: Use relay contacts 1 and 3 for beam-on operation; use 2 and 3 for beam-off operation.

3.4 TRANSISTORIZED SILICON-CELL RELAY (LOW-CURRENT RELAY)

In the same way that the previous circuit boosts the output of a silicon photocell to operate a 10-mA relay, a similar transistorized dc amplifier may be used with a 1-mA relay to sensitize a given photocell significantly (Fig. 3-4).

Here, a single 2N190 transistor (Q) in a common-emitter circuit amplifies the output of the S1M-C photocell (PC) to operate a 5F-1000S 1-mA, 1000-ohm relay, RY. The photocell must be polarized so that its negative (black) output lead is connected to the base of the transistor. A 6-V battery (B) sup-

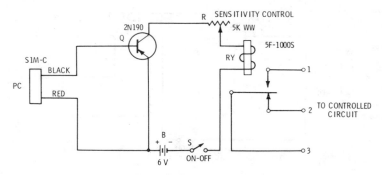

Fig. 3-4. Transistorized silicon-cell relay (1-mA relay).

plies the required dc power. An illumination of approximately 30 footcandles closes the relay.

The relay contacts will handle 0.25 ampere. This current rating will allow some devices to be operated directly from the relay. In other instances, the light-operated relay may be used to operate an external heavier-duty relay. Operation may be obtained with the light beam on or off: Use relay contacts 1 and 3 for beam-on operation; use 2 and 3 for beam-off operation.

3.5 TRANSISTORIZED SELENIUM-CELL RELAY (12-V SUPPLY)

The comparatively low output of a small selenium photocell may be boosted, by means of a transistorized dc amplifier, to operate a relay. Figs. 3-5 and 3-6 show suitable circuits. The first uses a 12-V battery and 2-mA relay, and is described in this section. The second employs a 3-V battery and 1-mA relay, and is described in Section 3.6. In either instance, the photocell could not drive the relay directly.

In Fig. 3-5, a single 2N190 transistor (Q) in a common-emitter circuit amplifies the output of a B3M-C selenium photocell (PC) to operate a 4F-5000S 2-mA, 5000-ohm relay (RY). The photocell must be polarized so that its negative (black) output lead is connected to the base of the transistor. A 12-V battery (B) supplies the required dc power. An illumination of approximately 100 footcandles closes the relay.

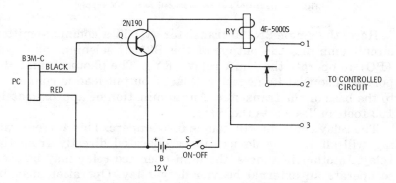

Fig. 3-5. Transistorized selenium-cell relay (12-V supply).

The relay contacts will handle 2 amperes. This current will allow most devices to be operated directly from the relay. In other instances, the light-operated relay may be used to operate an external heavier-duty relay. Operation may be obtained with the light beam on or off: Use relay contacts 1 and 3 for beam-on operation; use 2 and 3 for beam-off operation.

3.6 TRANSISTORIZED SELENIUM-CELL RELAY (3-V SUPPLY)

The circuit shown in Fig. 3-6 is similar to the one shown in Fig. 3-5 and described in Section 3.5, except that a more sensitive relay, RY (type 5F-1000S 1-mA, 1000-ohm), is used, and this allows reduction of battery (B) voltage to 3 volts.

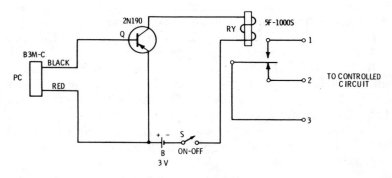

Fig. 3-6. Transistorized selenium-cell relay (3-V supply).

Here, the single 2N190 transistor (Q) in a common-emitter circuit amplifies the output of the B3M-C selenium photocell (PC) to operate the 1-mA relay (RY). The photocell must be polarized so that its negative (black) output lead is connected to the base of the transistor. An illumination of approximately 100 footcandles closes the relay.

The relay contacts will handle 0.25 ampere. This current rating will allow some devices to be operated directly from the relay. In other instances, the light-operated relay may be used to operate an external heavier-duty relay. Operation may be obtained with the light beam on or off: Use relay contacts 1 and 3 for beam-on operation; use 2 and 3 for beam-off operation.

3.7 IC-TYPE DC PHOTOCELL AMPLIFIER

Fig. 3-7 shows the circuit for a simple integrated-circuit dc amplifier which can boost the output of a photocell used in control circuits. This circuit employs only two outboard components in addition to the photocell: sensitivity-control potentiometer R_1 and negative-feedback resistor R_2.

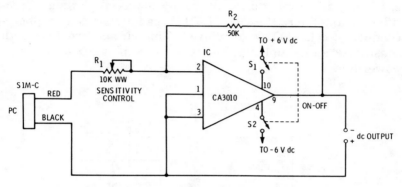

Fig. 3-7. IC-type dc photocell amplifier.

The type CA3010 integrated circuit operated in this configuration gives a dc voltage gain of 4 when R_1 is set to its lowest resistance (maximum amplifier sensitivity). The S1M-C photocell (PC), a silicon unit, is connected for positive dc output (red lead) to the inverting terminal (2) of the IC. In typical operation, with R_1 set for maximum sensitivity, bright sunlight or equivalent artificial illumination gives 2 volts (open circuit) at the DC OUTPUT terminals.

For increased amplification of the photocell voltage (increased sensitivity of the system), two or more identical dc amplifier stages may be operated in cascade. Other ICs will give similar performance; however, the experimenter must keep in mind that the maximum output-voltage swing is determined by the IC output characteristics and the amount of negative feedback introduced by R_2. Cascading will increase the overall sensitivity, making it possible to obtain a given dc output voltage with a lower level of illumination, but cascading will not increase the maximum obtainable dc output voltage for a given IC.

The circuit of Fig. 3-7 may be used ahead of most dc-responsive, high-resistance-input control and counting devices, oscilloscopes, graphic recorders, and similar equipment.

3.8 CADMIUM-SELENIDE-CELL AC RELAY

The arrangement shown in Fig. 3-8 makes possible the operation of an ac relay directly from a photocell, without a rectifier or amplifier. Here, a CL5M4L cadmium-selenide photoconductive cell (PC) is employed. This circuit provides both sensitivity and simplicity, but it must be plugged into the ac power line.

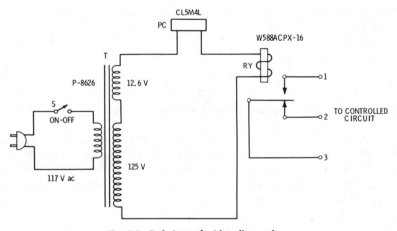

Fig. 3-8. Cadmium-selenide-cell ac relay.

The P-8626 transformer (T) applies 137.6 volts ac to the photocell and ac relay, RY (type W588ACPX-16, 115-V, 2200-ohm), in series. This voltage, the exact value of which is not needed, is obtained by connecting the 125-V and 12.6-V secondary windings in series-aiding. While this is a convenient arrangement with the readily available transformer shown here, any single-secondary transformer that will supply approximately 130 V at 60 mA also can be used.

The photoconductive cell acts as a light-variable resistor in series with the transformer and relay. An illumination of approximately 2 footcandles closes the relay.

The relay contacts will handle 10 amperes. This current rating will allow most devices to be operated directly from the relay. In the few other instances, the light-operated relay may be used to operate an external heavier-duty relay. Operation may be obtained with the light beam on or off: Use relay contacts 1 and 3 for beam-on operation; use 2 and 3 for beam-off operation.

Special ac relays may be used in this circuit, instead of the simple one shown. These include latching relays, ratchet relays, and stepping relays.

3.9 CHOPPED-LIGHT RELAY

Some photoelectric-relay applications require that the light beam be either modulated in intensity or chopped at an audio frequency, commonly 400 or 1000 Hz. Fig. 3-9 shows a relay circuit that utilizes such a beam.

In this arrangement, the ac (modulation) component in the output of an S1M-C silicon photocell (PC) is applied, through coupling capacitor C_1, to the input of a CA3010 integrated-circuit amplifier (IC). This IC is powered by a dual supply: posi-

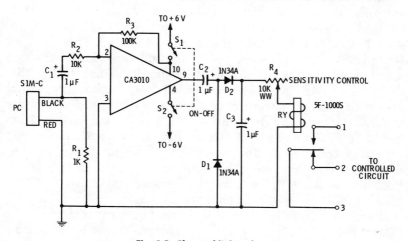

Fig. 3-9. Chopped-light relay.

tive 6 V and negative 6 V (this can be two batteries). The amplified ac output of the IC is rectified by the two 1N34A diodes (D_1 and D_2), and the resulting dc operates relay RY (type 5F-1000S 1-mA, 1000-ohm). Capacitor C_3 not only boosts the relay current, but also prevents relay chatter at the audio frequency. All three capacitors (C_1, C_2, C_3) are low-voltage, transistor-circuit-type electrolytic units. A modulated or chopped illumination of approximately 20 footcandles will cause the relay to close.

The relay contacts will handle 0.25 ampere. This current rating will allow some devices to be operated directly from the relay. In some instances, the light-operated relay may be used to operate an external heavier-duty relay. Operation may be obtained with the modulated, or chopped, beam on or off: Use relay contacts 1 and 3 for beam-on operation; use 2 and 3 for beam-off operation.

3.10 SHARPLY TUNED MODULATED-BEAM RELAY

The relay circuit shown in Fig. 3-10 tunes sharply—by adjustment of the single 5K wirewound potentiometer, R_4—from 200 to 2000 Hz. Thus, it may be tuned "on the nose" to the frequency of a modulated light beam.

The circuit is built around a single CA3140 operational-amplifier IC. The selectivity is obtained by means of a tunable bridged-differentiator RC null circuit (R_3-C_2-C_3-C_4-R_4) inserted into the negative-feedback loop of the IC. This network feeds back all frequencies except its null frequency, which is selected by adjustment of potentiometer R_4. The result is that the gain of the IC is cancelled on all frequencies except that one, which the IC accordingly transmits with good selectivity and little loss. A single rotation of R_4 tunes the circuit from 200 to 2000 Hz, and a dial attached to this potentiometer may be graduated directly in hertz. The IC output is rectified by diodes D_1 and D_2, and the resulting dc operates the 1-mA, 1000-ohm relay, RY. The circuit draws approximately 3.6 mA from each of the 9-volt dc sources (B_1, B_2).

A reader who is interested in tuning to frequencies outside of the 200–2000-Hz range may accomplish this by switching capacitors C_2, C_3, and C_4 in matched threes. Thus, when $C_2 =$

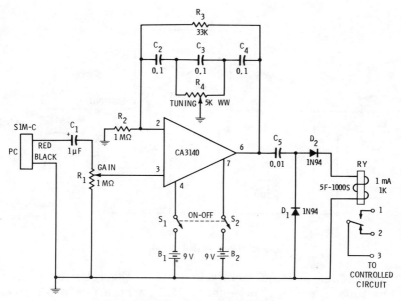

Fig. 3-10. Sharply tuned modulated-beam relay.

$C_3 = C_4 = 1 \ \mu F$, the range becomes 20 to 200 Hz; and when $C_2 = C_3 = C_4 = 0.01 \ \mu F$, the range becomes 2 to 20 kHz.

Operation may be obtained with the light beam either on or off: Use relay contacts 1 and 3 for beam-on operation; use 2 and 3 for beam-off operation.

3.11 LIGHT-OPERATED TIME-DELAY RELAY

Fig. 3-11 shows the circuit of an IC-type relay which holds-in for an adjustable time interval following the momentary application of a light beam. In this circuit, the CA3140 operational-amplifier IC is operated as a high-input-impedance dc follower (output terminal 6 connected directly to inverting-input terminal 2). When the S7M-C photocell (PC) is momentarily illuminated, its dc output charges the 100-μF electrolytic capacitor, C_1. At the same time, the positive output voltage of the cell is applied to input terminal 3 of the IC, and the relay accordingly picks up. The capacitor charge slowly leaks out through the timing resistor, R_1, and when this charge

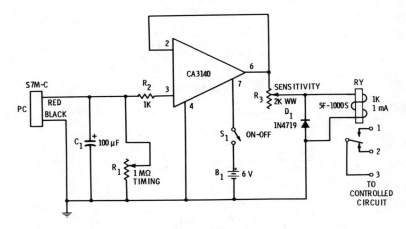

Fig. 3-11. Light-operated time-delay relay.

voltage falls to a sufficiently low value, the relay drops out. The time interval, t, is equal to $R_1 C_1$, where t is in seconds, R_1 in ohms, and C_1 in farads. With the 100-μF and 1-megohm values shown here, the maximum interval is 100 seconds. Longer intervals are obtainable by increasing C_1, R_1, or both.

Rheostat R_3 allows the sensitivity of the circuit to be adjusted with an individual light source, to insure that relay RY responds positively.

The output terminals may be used either to close or to open a circuit. Use relay contacts 1 and 3 to close the circuit; use 2 and 3 to open the circuit.

3.12 HEAVY-DUTY CONTROL CIRCUIT

Fig. 3-12 shows a heavy-duty circuit which can perform substantial work where dc operating current up to 1 ampere can be switched or varied photoelectrically without the intermediary of an electromechanical relay. An SK3009 power transistor (Q) is used to boost the output of an S7M-C silicon photocell (PC). The photocell must be polarized so that its negative (black) output lead is connected to the base of the transistor.

The load device, operated directly in the collector circuit of the transistor, must have low resistance (maximum, 25 ohms) and may be a solenoid, motor, electromechanical counter, actuator, stepping switch, or similar device. Power is supplied by a 6-volt storage battery or by an equivalent power-line-operated dc supply.

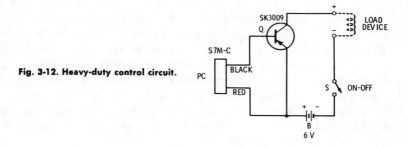

Fig. 3-12. Heavy-duty control circuit.

Light from a 40-watt incandescent lamp placed 1 foot from the photocell will cause approximately 0.25 ampere to flow through the load device; more intense light will drive the current up to 1 ampere. At current levels above 0.25 A, the transistor must be equipped with a heat sink.

When the photocell is darkened, the static collector current (I_{CO}) of the transistor, flowing through the load device, is 1 mA or less.

3.13 LIGHT-CONTROLLED SCR

Fig. 3-13 shows the circuit of a silicon controlled rectifier (SCR) which can be switched on by means of light. The SCR, like a thyratron tube, continues to conduct once it has been switched on, and will do so until its anode-supply voltage is

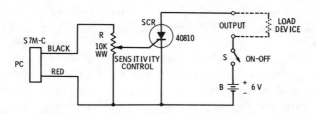

Fig. 3-13. Light-controlled SCR.

momentarily interrupted. Breaking the light beam, like interrupting the grid voltage of the thyratron, has no effect on the conduction, once started. The SCR thus acts as a fully electronic latched-in relay or switch.

In this arrangement, when the S7M-C silicon photocell (PC) is darkened and switch S is closed, little or no current flows through the load device connected to the output terminals. But when the photocell is illuminated, its resulting dc output is applied as a trigger voltage to the gate of the SCR. The SCR accordingly "fires," and a continuous current of the order of 2 amperes flows through the load device. This current continues to flow after the light is shut off and can be stopped only by opening switch S momentarily. The photocell is polarized so that its negative (black) output lead is connected to the SCR gate circuit through potentiometer R.

The load device may be any one of a variety of components, such as a motor, solenoid, actuator, lamp, alarm unit (bell, buzzer, horn, siren, etc.), or heater. For best results, the dc resistance of the load device should be 1 ohm or less. For higher-resistance devices, the voltage of the battery should be increased proportionately.

With the 10K wirewound potentiometer (R) set for maximum output, light from a 40-watt incandescent lamp placed 1 foot from the photocell will cause approximately 2 amperes to flow through the load device. The potentiometer acts as a sensitivity control for the circuit.

3.14 ALARM DEVICES: GENERAL NOTES

The photocell has a long history of association with burglar alarms and entry signals, and it is this application which, next to door openers, most often comes to the public mind when the electric eye is mentioned. The class name of these devices, regardless of the specific function they perform, is *intrusion alarms*.

Fig. 3-14 shows the basic arrangement of an intrusion alarm. In this system, a suitable light source (A) activates a photocell (B) which, in turn, operates a relay or relay circuit (C) which switches current from a battery or power supply (D) through an alarm (E).

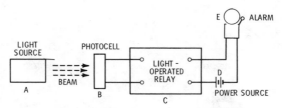

Fig. 3-14. Intrusion alarm—basic arrangement.

Any one of the photocell/relay combinations shown in Figs. 3-1 to 3-6 and 3-8 to 3-13 may be used directly as block C. Item E may be a bell, buzzer, horn, siren, lamp, or any other electrically operated attention catcher. For an ac-operated alarm, substitute an ac power source for the battery.

Most intrusion alarms operate by sounding off when the light beam is interrupted by the intruder. This means that in any of the relay circuits shown previously in this chapter, relay terminals 1 and 3 must be used. Also, in most instances it is desired that once the alarm is set off it continue to operate until it is reset by the owner. This necessitates that the electromechanical relay employed in the circuit be a latch-in type or—in lieu of this—that a fully electronic latch-in circuit, such as that shown in Fig. 3-13, be used.

Some circuits, such as those shown in Figs. 3-8 and 3-13, will operate a high-current alarm by themselves. Others, in which the relay contacts have relatively low current-handling capacity, must be used to close an external heavier-duty relay which, in turn, will operate the alarm.

For more comprehensive information on intrusion alarms, see the following books published by Howard W. Sams & Co., Inc.: *Building & Installing Electronic Intrusion Alarms* and *Security Electronics,* both by John E. Cunningham.

chapter **4**

Communications Circuits

This chapter describes 11 photoelectric circuits for use in various kinds of communications—light-beam, wire, and radio. The ones shown here will suggest others.

In these circuits, except where noted otherwise, resistances are specified in ohms, and capacitances in microfarads; resistors are 0.5 watt, and potentiometers are wirewound. In those circuits requiring a local dc supply, batteries are shown for simplicity; however, well-filtered, power-line-operated supplies may be used instead.

None of the devices demands a critical layout, so the reader is free to use his preferred method of assembly, such as breadboard, metal chassis, printed circuit, and so on. Only the electronics of the device is given here, mechanical details being left to the individual builder. For names and locations of manufacturers of the photocells and other special components shown in this chapter, see Appendix B.

The reader should study Section 4.1 before proceeding with the construction of any of the light-beam transmitters (Figs. 4-3 and 4-7) or light-beam receivers (Figs. 4-1, 4-2, 4-5, and 4-6).

4.1 LIGHT-BEAM COMMUNICATORS: PRELIMINARY REMARKS

The effectiveness of communication devices that use a light beam to link the transmitting station and the receiving station depends upon the intensity of the light source and whether or not lenses and/or reflectors are employed. This applies equally to the light-beam transmitters (Figs. 4-3 and 4-7) and the light-beam receivers (Figs. 4-1, 4-2, 4-5, and 4-6). The practical operating distance, as well as the strength of the signal, depends upon the amount of light that can be projected from the transmitter to the receiver. An important factor, too, is how well the receiving photocell can be protected from the ambient light.

Novelty houses, government-surplus stores, hobby shops, and some electronics suppliers carry various reflectors and lens systems, hoods, and shields which may be used to concentrate the light beam, and these should be chosen to serve the individual reader's requirements. In some cases, it will be desirable to have a reflector at the transmitter and a lens system at the receiver; in other cases, the reverse will be true. In still other instances, it will be best to have lenses at both ends, or reflectors at both ends. Sometimes, a color filter at each end will minimize the effects of interference from light other than that of the beam.

The transmitting lamp and the receiving photocell, together with any optical adjuncts, must be solidly supported so that they remain aligned with each other. Moreover, both must be free from vibration which would modulate the light beam at either the transmitting or receiving end. In many cases, the lamp and photocell will be mounted outdoors, and this calls for protection against weather. Usually, such protection will consist of enclosure in a waterproof case with a glass window which needs periodic cleaning. When the receiver is operated indoors, the photocell usually needs protection only from dust, and occasionally from vapor or fumes.

Careful building and installation of a light-beam communicator will contribute a great deal to the successful performance of the system. Sturdy construction practices are strongly recommended.

4.2 SIMPLE LIGHT-BEAM RECEIVER

Fig. 4-1 shows the circuit of a rudimentary, light-beam receiver which may be used for either voice or telegraphy. It consists of a type S7M-C silicon photovoltaic cell (PC) connected in series with either a pair of 2000-ohm magnetic headphones or an equivalent earpiece.

When the light beam impinges upon the photocell, the latter generates a voltage which actuates the headphones. This beam may be keyed and come from a simple light-beam telegraphy transmitter (see Fig. 4-7), whereupon it will produce Morse-type dot-and-dash clicks in the headphones, or it may be voice modulated and come from a suitable modulated-light-beam transmitter (see Fig. 4-3). In either case, the stronger the beam, the louder the signal.

A particular advantage of this simple receiver is its ability to operate without battery, power supply, or amplifier. The polarity of the photocell is unimportant.

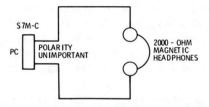

Fig. 4-1. Simple light-beam receiver.

4.3 SENSITIVE LIGHT-BEAM RECEIVER

For a louder signal, the output of the simple receiver described in Section 4.2 may be boosted with an audio amplifier. This results in the circuit shown in Fig. 4-2. Here, the output of the S7M-C photocell (PC) is amplified by a CA3020 integrated circuit (IC). This IC delivers a stout headphone signal; it also supplies 0.25 watt to a 3.2-ohm speaker when a 9-V battery (B) is employed. Slightly less output power is obtained when B is 6 volts. A heat sink should be used with the IC.

The tiny IC contains four direct-coupled amplifier stages, the last one being push-pull. Eight outboard components are required (C_1, C_2, C_3, C_4, R_1, R_2, R_3, and T). The output transformer (T) is a miniature, transistor-type unit having a 125-

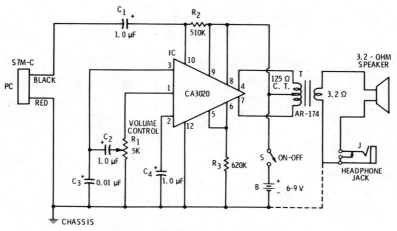

Fig. 4-2. Sensitive light-beam receiver.

ohm, center-tapped primary and 3.2-ohm secondary. The closed-circuit jack (J) allows 2000-ohm magnetic headphones (or equivalent earpiece) to be plugged-in in series with the voice coil of the speaker. This results in a substantial headphone signal, but the high resistance of the headphones effectively disables the speaker.

The light beam which is received by photocell PC may be keyed and come from a simple light-beam telegraph transmitter (see Fig. 4-7), whereupon it will produce Morse-type dot-and-dash clicks in the headphones or speaker; or it may be voice modulated and come from a suitable modulated-light-beam transmitter (see Fig. 4-3).

4.4 MODULATED-LIGHT-BEAM TRANSMITTER

Fig. 4-3 shows the circuit of a light-beam transmitter suitable for the transmission of voice signals or tone-type (mcw) telegraph signals. In this arrangement, a microphone (for voice) or a keyed audio oscillator (for tone telegraphy) is plugged into input jack J. A CA3020 integrated circuit (IC) amplifies the signal from the microphone or oscillator, and the 0.25-watt output of the IC fluctuates the brilliance of the lamp (L).

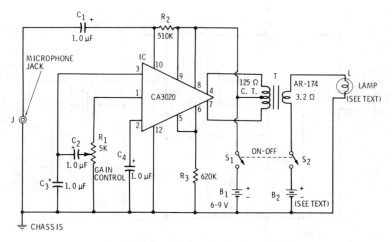

Fig. 4-3. Light-beam transmitter.

Battery B_1 supplies dc operating power to the IC. When the voltage of B_1 is 9 V, the IC output is 0.25 watt and the maximum-signal current drain is approximately 85 mA (at this level, the IC requires heat sinking). Battery B_2 supplies voltage to the lamp, and its rating will depend upon the type of lamp used. For many light-duty communications, a type 112, lens-end flashlight lamp will suffice and B_2 will be 1.5 V. For heavier-duty work, a sealed-beam, automobile-type lamp can be used, and B_2 will be 6 V or 12 V. In either case, gain control R_1 must be set so that modulation will not brighten the lamp to the extent of burning it out.

For full 0.25-watt undistorted output from the IC, an input-signal amplitude of 50 mV rms (from microphone or oscillator) is required at jack J when gain control R_1 is set for maximum gain. Audio power is delivered to the lamp through T, a miniature, transistor-type output transformer having a 125-ohm center-tapped primary and a 3.2-ohm secondary.

4.5 SUN-POWERED MORSE WIRE TELEGRAPH

A simple wire telegraph of a Morse type (clicking dots and dashes) is easily powered with a sunlit photovoltaic cell. Fig. 4-4 shows this arrangement.

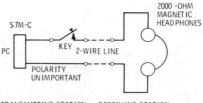

Fig. 4-4. Sun-powered Morse telegraph.

TRANSMITTING STATION   RECEIVING STATION

Here, a single S7M-C silicon photocell (PC) delivers up to 3 V in bright sunlight and acts as a simple solar battery. The polarity of the cell is unimportant. The dc output of the cell is broken up into long and short intervals (dashes and dots) by means of the key, and these pulses are fed over a two-wire line to a distant pair of 2000-ohm magnetic headphones or an equivalent earpiece. A single-wire line also may be used if a good ground is available for the return circuit.

For two-way communication and to prevent hearing one's own signals, the circuit may be duplicated and reversed, that is, with the photocell and key of the second circuit at the first receiving station, and the headphones of the second circuit at the first transmitting station. In this way, each station will have a key and headphones.

4.6 SENSITIVE LIGHT-BEAM MORSE TELEGRAPH RECEIVER

The light beam from a simple transmitter is easily keyed, by switching the lamp on and off, to form dot-and-dash telegraph signals. Such a transmitter is shown in Fig. 4-7. These signals can be received as clicks with the simple circuit shown in Fig. 4-1. A more sensitive arrangement, however, is shown in Fig. 4-5. This circuit is capable of operating over longer distances than those afforded by the simple circuit, and it gives either headphone or speaker clicks.

In this setup, the keyed light beam is picked up by the sensitive, cadmium-sulfide-photocell relay shown earlier in Fig. 3-2, Chapter 3, and used in its entirety here. This relay is closed by an illumination of approximately 5 footcandles and, in turn (in Fig. 4-5), switches the current from 7.5-V battery B through either the headphones or the speaker-coupling trans-

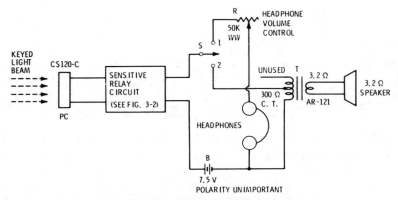

Fig. 4-5. Sensitive light-beam Morse telegraph receiver.

former (T). When switch S is thrown to position 1, headphone operation is obtained; and when S is thrown to position 2, speaker operation is obtained. Relay contacts 1 and 3 are used (see Fig. 3-2, Chapter 3).

The coupling transformer is a miniature, transistor-type output unit having a 300-ohm, center-tapped primary and 3.2-ohm secondary.

4.7 LIGHT-BEAM TONE TELEGRAPH RECEIVER

The last two circuits (Sections 4.5 and 4.6) utilize the American Morse code (click-type dots and dashes). Fig. 4-6, however, shows a circuit for the Continental Morse code (tone-type dots and dashes) which is standard for radio telegraphy and for buzzer or oscillator wire telegraphy. In this arrangement, a single-frequency audio oscillator is powered by an S7M-C silicon photocell (PC). The cell is actuated by the received, keyed light beam, so the oscillator has no dc power and cannot operate except when the cell is illuminated. The keyed beam may be generated by the simple light-beam transmitter shown in Fig. 4-7.

The oscillator employs an inexpensive, 2N2712 silicon transistor (Q) in a common-emitter, transformer-feedback circuit. The transformer (T) is a miniature, transistor-type, interstage coupling unit having a 10,000-ohm primary and a 2000-ohm secondary. It provides regenerative feedback between the

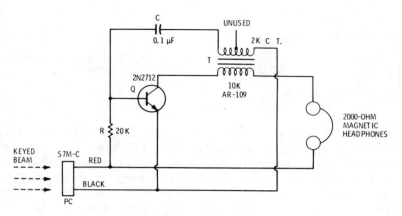

Fig. 4-6. Light-beam tone-telegraph receiver.

collector output circuit and the base input circuit of the transistor. The windings must be correctly polarized for regeneration; if oscillation does not occur, reverse the connections on one winding only. The oscillation frequency is determined by the inductance of the full secondary winding and the capacitance of coupling capacitor C. With the 0.1 μF shown and the model of transformer specified, the frequency is approximately 480 Hz. For higher frequencies, reduce the capacitance below 0.1 μF; for lower frequencies, increase the capacitance above this figure.

A particular advantage of this circuit is the location of the tone generator at the receiving station. This allows use of an unmodulated, merely keyed, light beam which can be obtained with a very simple transmitter, such as the one shown in Fig. 4-7. Furthermore, the choice of tone frequency is entirely up to the receiving operator.

4.8 LIGHT-BEAM TELEGRAPH TRANSMITTER

In the simplest form of light-beam telegraph transmitter, a lamp is merely switched on and off, with a telegraph key, to form the dots and dashes. This arrangement is shown in Fig. 4-7.

The lamp should be provided with a lens system or a reflector, as desired (see Section 4.1). For many light-duty commu-

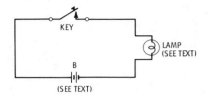

Fig. 4-7. Light-beam telegraph transmitter.

nications, a type 112 lens-end flashlight lamp will suffice, and for this lamp, battery B will be 1.5 V. For heavier-duty work, a sealed-beam, automobile-type lamp is serviceable, and B will be 6 V or 12 V. If a 115-V lamp is used, it should not be switched directly with the key, but through a suitable relay, to guard against electric shock and to protect the key contacts.

This simple transmitter may be used with the receivers shown in Figs. 4-1, 4-2, 4-5, and 4-6.

4.9 SUN-POWERED WIRE TELEPHONE

A simple wire telephone is easily powered with a sunlit photovoltaic cell. Fig. 4-8 shows this arrangement. Here, a single S7M-C silicon photocell (PC) delivers up to 3 V in bright sunlight and acts as a simple solar battery. The polarity of the cell is unimportant.

Single-pole, double-throw switches (S_1 at Station A, and S_2 at Station B) enable either station to talk or listen, as desired. For the simple series circuit shown here, carbon microphones and 2000-ohm magnetic headphones (or equivalent earpieces)

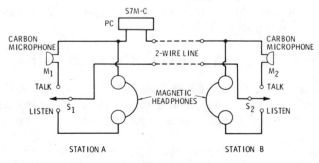

Fig. 4-8. Sun-powered wire telephone.

give the best results. For increased volume, two photocells may be conneced in series aiding.

While a two-wire line will be best in most cases, a single-wire line also may be used if a good ground is available for the return circuit.

4.10 SUN-POWERED TRANSMITTER FOR RADIO AMATEURS

Fig. 4-9 shows the circuit of a low-powered cw transmitter which, with a good antenna and a clear frequency, can give a surprisingly good account of itself. It consists of a crystal oscillator embodying a 2N3823 field-effect transistor (Q). The dc drain power of the transistor is supplied by three S7M-C silicon photocells (PC_1, PC_2, PC_3) connected in series to form a solar battery. In bright sunlight, or equivalent artificial illumination, these cells deliver up to 3 V each.

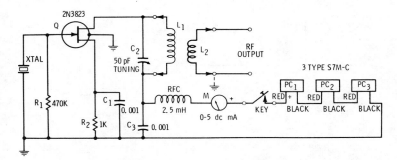

Fig. 4-9. Sun-powered cw radio transmitter.

Standard amateur-band plug-in coils (L_1 and L_2) may be used for tuning. The crystal (XTAL) is selected for the desired operating frequency. Tuning of the transmitter is conventional:

1. Expose the photocells to bright sunlight or equivalent artificial illumination.
2. With the antenna temporarily disconnected from the RF OUTPUT terminals and the key depressed, adjust variable capacitor C_2 until coil L_1 is tuned to the crystal frequency,

as indicated by a sharp dip in the drain current indicated by milliammeter M.

3. Connect the antenna, noting that the meter reading rises.
4. Retune C_2 to ensure that the circuit is still adjusted to drain-current dip, as noted in Step 2.

4.11 SUN-POWERED NONREGENERATIVE BROADCAST RECEIVER

Fig. 4-10 shows the circuit of a simple standard-broadcast-band radio receiver powered by three S7M-C silicon photocells (PC_1, PC_2, PC_3) connected in series to form a solar battery. The receiver employs a 1N34A diode detector (D_1) and a 2N190 transistor (Q) audio amplifier.

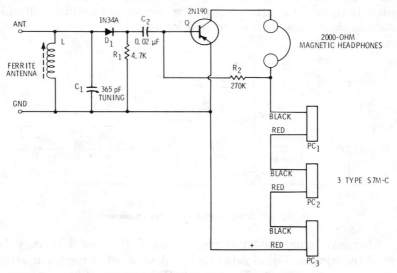

Fig. 4-10. Sun-powered nonregenerative broadcast receiver.

Inductor L is a conventional ferrite-rod am antenna, and this component is tuned by means of the 365-pF variable capacitor (C_1) to cover the broadcast band. Close alignment may be effected by means of a one-time adjustment of the "screwdriver-tuned" core of the antenna. Strong, nearby stations may be picked up directly by this antenna; others require an external antenna and ground, connected to the ANT and GND terminals, respectively.

The receiver gives a substantial signal in the 2000-ohm magnetic headphones (or an equivalent earpiece), since the three photocells connected in series-aiding yield up to 3 V each when exposed to bright sunlight or equivalent artificial illumination.

4.12 SUN-POWERED REGENERATIVE BROADCAST RECEIVER

Considerably better sensitivity than is possible with the receiver circuit described in Section 4.11 may be obtained by means of regeneration. Fig. 4-11 shows such a regenerative circuit.

In this arrangement, a 2N2608 field-effect transistor (Q) is employed in a Hartley-type circuit. The heart of this circuit is a tapped ferrite-rod am antenna (L), which serves as both the internal antenna and the oscillator coil, and is tuned by means of the 365-pF variable capacitor (C_2). A 200-ohm rheostat (R_2) permits adjustment of the regeneration intensity and therefore the sensitivity of the receiver before oscillation

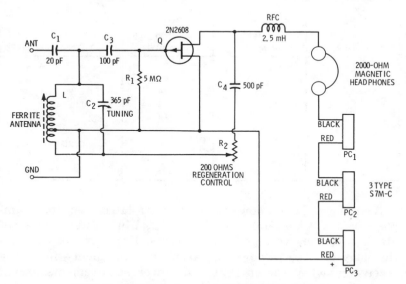

Fig. 4-11. Sun-powered regenerative broadcast receiver.

(squealing) sets in. As in the previous circuit, dc power is supplied by three S7M-C silicon photocells (PC_1, PC_2, PC_3) connected in series to form a solar battery. These cells can yield up to 3 V each when exposed to bright sunlight or equivalent artificial illumination.

Most broadcast stations within reasonable range of the receiver can be picked up by the ferrite-rod antenna alone; others require an external antenna and ground, connected to the ANT and GND terminals, respectively.

4.13 SUN-POWERED STANDARD RECEIVERS

Simple, one-transistor receivers are shown in Figs. 4-10 and 4-11. This, however, is not the full extent to which sun power may be applied to radio reception. A full-sized portable transistor radio, for example, may be powered by using the required number of power-type silicon modules in series and/or parallel to supply the desired operating voltage and current.

Table 4-1. Power Solar Module Data

Output Voltage (min V @ min mA)	Output Current (min mA)	Output Power (min mW)	Type Number
0.4	60	24	S2900E5M
0.4	90	36	S2900E7M
0.4	120	48	S2900E9.5M
1.6	36	58	SP2A40B
1.6	40	64	SP2B48B
1.6	72	115	SP2C80B
1.6	80	128	SP2D96B
3.2	36	115	SP4C40B
3.2	40	128	SP4D48B

A number of such power units are available (see, for example, Fig. 1-5, Chapter 1). Table 4-1 lists nine power-type modules for various voltages and currents. Units of this description may be used to power a variety of equipment—including receivers—when the current and power requirements exceed the limits of standard photocells.

A string of standard photocells also can be used when an ample number of cells are available to the user. However, the internal resistance of a series string is the total resistance of the cells, and this increased resistance may degrade the voltage regulation of the resulting solar battery. Also, the parallel connection sometimes is not entirely successful unless the cells are identical in performance. Moreover, when there are many cells in a combination, some difficulty may be experienced in illuminating all of them equally.

chapter **5**

Miscellaneous Circuits and Devices

This chapter presents a group of photoelectric circuits and devices which do not fit into the categories established by the previous chapters. The number of areas of application included in this chapter, added to those in Chapters 2 to 4, point out the versatility of the photocell and solar cell.

In the circuits given in this chapter, except where noted otherwise, resistance is specified in ohms, and capacitance in microfarads; resistors are 0.5 watt, and potentiometers are wirewound. In those circuits requiring a local dc supply, batteries are shown for simplicity; however, well-filtered, power-line-operated supplies may be used instead. For names and locations of manufacturers of the photocells and other special components shown in this chapter, see Appendix B.

5.1 OBJECT COUNTERS

Opaque objects passing between a light source and a photocell interrupt the beam, temporarily deactivating the cell. If the cell is connected to a suitable totalizing counter, the latter will indicate the number of objects that have passed.

Fig. 5-1 shows a photoelectric counter setup with an electro-mechanical readout, the latter being a small, 4-dial, 117-V ac-

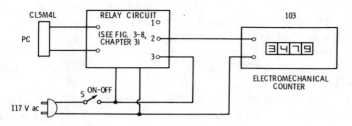

Fig. 5-1. Electromechanical-readout object counter.

type (Model 103) unit. The indicator is operated by the fully ac-operated relay which employs a CL5M4L cadmium-selenide photocell, shown in Fig. 3-8 and described in Section 3.8, Chapter 3. For this application (beam-off operation), relay contacts 2 and 3 must be used.

Since an illumination of approximately 2 footcandles actuates the relay, the light source need not be very bright. Each object passing between the light source and the photocell interrupts the beam, closes the number 2 and 3 contacts of the relay, and sends an ac pulse through the counter, advancing the readout one count.

Fig. 5-2 shows a fully electronic object counter, recommended when complete freedom from moving parts is desired. In this arrangement, the light sensor is a CS120-C cadmium-sulfide photoconductive cell (PC) in a 4-arm bridge circuit comprising resistor R_1, the two "halves" of the zero-set poten-

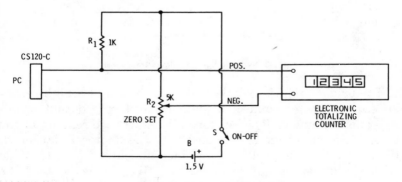

Fig. 5-2. Electronic-readout object counter.

tiometer (R_2) winding, and the internal resistance of the photocell. The bridge is powered by a 1.5-V cell.

With the photocell fully illuminated by the light source used in the setup, and a high-resistance dc voltmeter connected temporarily across the POS and NEG bridge output terminals, potentiometer R_2 is adjusted for null (exact zero deflection of the meter). Each time the cell is then darkened by an object passing in front of it, the bridge delivers a +1.15-V output pulse to the electronic counter. If a particular counter requires a negative pulse, simply reverse the battery connections.

5.2 VERSATILE LIGHT-TO-VOLTAGE TRANSDUCER

Fig. 5-3 shows a flexible circuit for translating light changes to voltage values. In this arrangement, a CS120-C cadmium-sulfide photoconductive cell (PC) operates in a four-arm bridge consisting of resistor R_1, the two "halves" of the potentiometer (R_2) winding, and the internal resistance of the photocell.

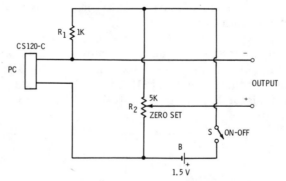

Fig. 5-3. Versatile light-to-voltage transducer.

The bridge is initially balanced with the aid of a high-resistance dc voltmeter connected temporarily to the output terminals; potentiometer R_2 is set for null (exact zero reading of the meter). If the bridge is balanced with the photocell darkened, illumination of the cell will produce an output of 1.15 V dc with the upper output terminal negative. Conversely, if the bridge is balanced with the photocell illuminated, darkening of the cell will produce an output of 1.15 V dc with the upper output terminal positive. For the opposite output polarity, reverse the output terminals or battery B. For higher output voltage, increase the battery voltage (the CS120-C photocell has maximum ratings of 20 V and 0.4 W).

5.3 LIGHT-OPERATED VOLTAGE DIVIDER (POTENTIOMETER)

A photoconductive cell, either cadmium sulfide or cadmium selenide, may be used as the variable-resistor section of a voltage divider. Fig. 5-4 shows such an arrangement, which will operate on ac, dc, or a mixture of the two.

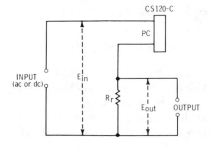

Fig. 5-4. Light-operated voltage divider (potentiometer).

The CS120-C cadmium-sulfide cell (PC) shown here has a specified dark resistance of 1600 ohms; at 100-footcandles (fc) illumination, the resistance falls to approximately 0.11 ohm. Thus, for a light variation of 100 fc, a resistance variation of more than 14,000 to 1 is obtained. Various other cadmium-sulfide and cadmium-selenide photoconductive cells offer other values of dark resistance and light resistance. The CL909L, for example, offers approximately 30 megohms of resistance at 0.01 fc, and approximately 3000 ohms at 100 fc.

When designing the voltage divider, choose resistance R_r for a satisfactory output value (in some applications, such as audio attenuators, 500 ohms is a common value). Next, determine the dark and light resistances of the photocell, either by means of measurements or by consulting the cell manufacturer's literature. Then, determine the voltage-division ratio afforded by the circuit from the relationship:

$$E_{out} = E_{in}[R_r/(R_c + R_r)]$$

where,

R_c is the photocell resistance,
R_r is the output-resistor value.

Solve the equation first for R_c equal to the dark resistance of the cell, and next for R_c equal to the light resistance of the cell.

The light-controlled voltage divider has a great many practical uses which readily come to the experimenter's mind. Among these is the *remote-controlled volume control* in which a small lamp illuminates the photocell in proportion to the setting of a rheostat at the master position. The lamp is mounted in a light-tight enclosure with the divider and is controlled from the master operating point. See also the discussion of optoelectronic couplers in Section 5.6.

5.4 LIGHT-CONTROLLED CAPACITOR

In Fig. 5-5, a varactor (D) receives its variable-dc control voltage from an S7M-C silicon photocell (PC) through the usual high-resistance isolating resistor (R) used in dc-controlled varactor circuits. The photocell shown here will vary the capacitance of the 1N4815A varactor from approximately 260 pF, when the cell is completely darkened, to approximately 115 pF, when the cell is receiving bright sunlight (or equivalent artificial illumination) and is delivering 3 V dc. A 0.01-μF mica or ceramic blocking capacitor (C) protects the varactor from any direct current present in the circuit in which this light-variable capacitor is used, and also prevents the external circuit from short-circuiting the varactor dc control voltage. The capacitance of C must be considerably higher than the

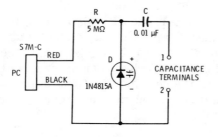

Fig. 5-5. Light-controlled capacitor.

maximum capacitance of the varactor, so that the capacitance "seen" at terminals 1 and 2 is principally that of the varactor. Because the varactor draws virtually no current from the photocell, the output voltage of the cell will attain its maximum no-load value at each level of illumination.

The variable capacitance provided by this circuit may be used in a variety of ways. For example, a telemetering transmitter may be tuned by it to communicate light-intensity values to a remote station. Or, it can be used to shift the frequency of a beat-frequency oscillator in response to the interruption or variation of illumination. In any application, the amplitude of an ac signal applied to the varactor by the external circuit must be very much lower than the photocell dc voltage, to prevent overriding the latter.

5.5 STABILIZING SOLAR-CELL OUTPUT

The dc output of a solar cell or solar battery sometimes varies widely under conditions of fluctuating illumination. To remedy this condition, automatic regulation of the output voltage may be obtained by operating either a voltage-dependent resistor (vdr) or a high-capacitance capacitor in parallel with the cell.

Fig. 5-6A shows the vdr circuit. Here, the nonlinear resistor (R) acts as a self-regulating element. During low illumination, the cell voltage is low and the vdr resistance is high. But during bright illumination, the cell voltage is high, which reduces the vdr resistance; the vdr then conducts heavily and pulls the voltage down. The net result of this action is to keep the output voltage constant in the face of fluctuating illumination. The

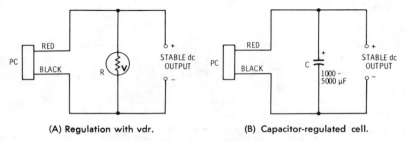

(A) Regulation with vdr. (B) Capacitor-regulated cell.

Fig. 5-6. Stabilized-output solar-cell circuits.

vdr selected for this application must show good nonlinearity at the photocell voltages.

Fig. 5-6B shows the capacitor circuit. Here, a high-capacitance, low-voltage, electrolytic capacitor, C (1000 to 5000 μF at 6 or 12 V), acts as the self-regulating element. When the cell is normally illuminated, its dc output charges the capacitor. When the light intensity falls, the large capacitance cannot discharge instantaneously, so the capacitor holds the dc output voltage to the original level for an interval, often long enough for the illumination to recover its initial level. Thus, the normally sluggish response of this capacitor produces a stabilized dc output.

For either of these circuits to be most effective, the resistance of the load device connected to the STABLE DC OUTPUT terminals must be high. For the vdr circuit (Fig. 5-6A), this prevents severe modification of the nonlinear curve of special resistor R; for the capacitor circuit (Fig. 5-6B), this prevents rapid discharge of the capacitor.

5.6 OPTOELECTRONIC COUPLER

A photocell and a light source may be enclosed inside a light-tight housing to provide a high-isolation signal coupler of a type which, in various forms, has found many uses in modern electronics. This device is shown in Fig. 5-7.

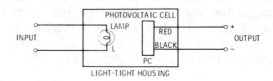

(A) Photovoltaic model.

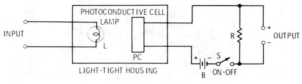

(B) Photoconductive model.

Fig. 5-7. Optoelectronic couplers.

In Fig. 5-7A, a small incandescent lamp (L) is mounted close to a photovoltaic cell (PC). The input-signal voltage lights the lamp to a brilliance proportional to this voltage; the lamp, in turn, activates the photocell. The cell then generates the output-signal voltage, which is proportional to the brilliance and thus to the input-signal voltage, but is not necessarily equal to the latter. Either an ac or a dc input signal, or a mixture of the two, may be used.

In Fig. 5-7B, a small incandescent lamp (L) is mounted close to a photoconductive cell (PC). The cell is biased by battery B. The input-signal voltage lights the lamp to a brilliance proportional to this voltage; the lamp, in turn, activates the photocell, causing its resistance to fall in proportion to the brilliance. This action causes a current to flow from battery B through output resistor R, and to develop the output-signal voltage across this resistor. Depending upon the voltage of battery B,

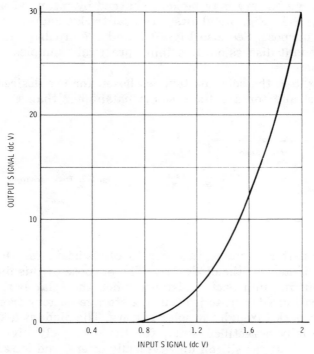

Fig. 5-8. Performance of cadmium-sulfide coupler.

the output-signal voltage may be equal to, lower than, or higher than the input-signal voltage. Thus, it is easily possible to obtain voltage amplification with the circuit in Fig. 5-7B. In this connection, Fig. 5-8 shows the performance of the circuit using a CS120-C photocell, a 2-V, 60-mA (Type 48) pilot lamp, B = 90 volts, and R = 10,000 ohms. Note from this curve that the voltage amplification varies with the input-signal voltage, being approximately 1 at 1-V input, and 30 at 2-V input.

It should be noted that light sources other than the incandescent lamp often are used in optoelectronic couplers. These include neon lamps and light-emitting diodes. (Neon lamps, of course, are on-off devices and cannot give a smooth variation of light intensity.)

5.7 TRICKLE CHARGER

A storage battery may be kept charged by means of a suitable solar cell. Fig. 5-9 shows a typical trickle-charger circuit for this purpose. See also Figs. 1-5 and 1-7 in Chapter 1 and the attendant discussion describing practical examples of this application.

In Fig. 5-9, the solar battery is chosen for the desired current level and for a voltage somewhat higher than the fully

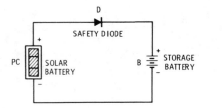

Fig. 5-9. Trickle charger.

charged battery voltage in order to compensate for the forward resistance of the diode (D). The purpose of this diode is to present its high back resistance when the solar battery is not operating (dark), to prevent the storage battery from discharging back through the solar battery. The diode is a silicon unit (usually a rectifier unit), for the extremely high back resistance that the silicon unit typically offers, and is rated to carry the desired charging current.

See Table 4-1, Chapter 4, for a list of power-type silicon solar modules that can be used—singly, in series, in parallel, or in series-parallel—for battery charging.

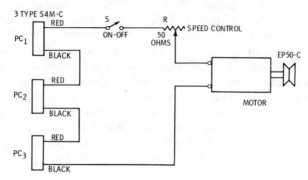

Fig. 5-10. Light-powered motor.

5.8 LIGHT-POWERED MOTOR

Fig. 5-10 shows a setup in which a low-friction-bearing dc motor is driven by a solar battery consisting of three S4M-C silicon photocells (PC_1, PC_2, PC_3) connected in series. When the cells are illuminated by bright sunlight (or equivalent artificial illumination), the small EP50-C motor develops sufficient torque to drive lightweight models and devices as attention catchers in science-fair exhibits, advertising displays, and demonstrations of sun power.

When the light has a fixed, steady intensity, the 50-ohm wirewound rheostat (R) may be used to vary the speed of the motor. Otherwise, R may be set to zero resistance and the speed varied by adjusting the intensity of the light.

Fig. 5-11. Slave photoflash.

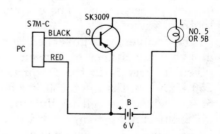

5.9 SLAVE PHOTOFLASH

Fig. 5-11 shows the simple circuit of a slave flash unit for photography. Light from the main flashlamp (in a camera-mounted gun, fastened to a wall or mounted on a tripod) acti-vates an S7M-C silicon photocell (PC). The resulting dc output voltage of the photocell biases the base of the SK3009 power transistor negatively and biases the emitter positively. This causes a momentary surge of current in the collector circuit of the transistor, and this surge fires the auxiliary flashbulb (L) which may be a Number 5 or 5B unit. Once the bulb fires, it blows out and opens the circuit between battery B and the col-lector of the transistor. Before the firing, only a tiny static collector current (I_{co}) flows through the bulb.

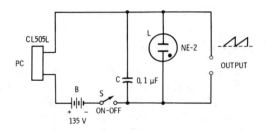

Fig. 5-12. Light-controlled neon oscillator.

5.10 LIGHT-CONTROLLED NEON OSCILLATOR

In Fig. 5-12, a CL505L cadmium-sulfide photoconductive cell functions as a light-variable, frequency-control resistor in a conventional neon-bulb relaxation oscillator circuit. The more intense the light impinging upon the cell, the higher the fre-quency of oscillation. The output signal is approximately saw-tooth in waveform.

The frequency is governed by the resistance of the photocell and the capacitance of capacitor C. With the circuit constants given in Fig. 5-12, the frequency varies from approximately 250 Hz when the photocell is completely darkened to well over 5 kHz when the cell is illuminated with 2 footcandles. For a different range, change the value of C (increasing the capaci-tance lowers the frequency, and vice versa).

5.11 OPTICAL TACHOMETER

When the rotating shaft or some other rotating or vibrating member of a machine periodically interrupts or reflects a light beam that strikes a photocell, the output of the cell is a pulsating voltage. The frequency of this voltage is proportional to the speed of the machine, and a tachometer can be obtained by applying this voltage to a simple audio frequency meter. Such an arrangement is shown in Fig. 5-13.

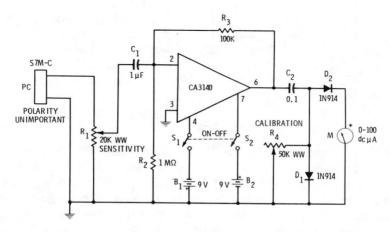

Fig. 5-13. Optical tachometer.

In this circuit, a CA3140 operational-amplifier IC is operated as a saturating amplifier. The amplifier output therefore is a constant-amplitude square wave. When this wave is presented to the meter circuit (C_2-D_1-D_2-M-R_4), the deflection of the meter is proportional only to the number of waves arriving each second. Thus, the IC plus the meter circuit constitutes a simple, linear-reading audio frequency meter whose input signal is supplied by the photocell.

The range of the instrument is governed by capacitance C_2 and the resistance setting of rheostat R_4. With the values shown in Fig. 5-13, the 0–100 dc microammeter (M) reads directly and linearly 0–100 Hz. This corresponds to 0–6000 rpm. Other full-scale values may be obtained by adjustment of rheostat R_4 when an audio generator is substituted temporarily for the photocell and tuned to a frequency in hertz equal to rpm/60.

5.12 COLOR MATCHER

A simple device for matching colors with enough accuracy for practical purposes is shown schematically in Fig. 5-14.

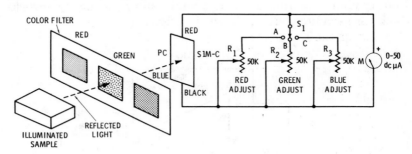

Fig. 5-14. Color matcher.

In this arrangement, the S1M-C silicon photocell (PC) views the color samples to be compared successively through red, green, and blue filters, and the deflection of the 0–50 dc microammeter (M) is recorded for each exposure. When the meter reading is the same for all three exposures (positions A, B, and C of switch S_1), the samples match in hue, saturation, and brilliance. The instrument operates on light reflected by the sample.

For positioning before the photocell, the color filter may be a slide, as shown in Fig. 5-14, or it may be a wheel. The transparent windows are made from Wratten gelatin sheet obtained in 2-inch by 2-inch sheets from a photographic supply house. The red is Eastman Kodak No. 25A, the green No. 58, and the blue No. 47.

When the instrument is in use, the sample must be illuminated with strong white light (sunlight is satisfactory), and the distance between sample and instrument must be held constant. Also, the cell must not receive light from any other source.

Use the color matcher in the following manner: (1) Place the standard sample into position and illuminate it. (2) Set switch S_1 to Position A. (3) Adjust RED rheostat R_1 for an M deflection near center scale. (4) Set switch S_1 to Position B. (5) Adjust GREEN rheostat R_2 for the same deflection as in Step 3. (6) Set switch S_1 to Position C. (7) Adjust BLUE rheostat R_3 for the same deflection as in Steps 3 and 5. (8) Replace the standard sample with the test sample. Take care that the latter is in the same spot previously occupied by the former and that the illumination is the same as before. (9) Set switch S_1 successively to its A, B, and C positions, observing the deflection of the meter for each. These readings will be identical with those taken with the standard sample if the two samples match exactly. When the second readings differ from the first, they indicate the relative excess or deficiency in the red, green, and blue components in the test sample.

appendix **A**

Glossary of Optical and Photoelectric Terms

A

absorption—Some of the light that strikes a surface passes into the material and is lost. This phenomenon is termed *absorption*.

angstrom—Abbreviated A, Å, or AU. A unit of extremely short wavelength. $1 \text{ Å} = 1 \times 10^{-8}$ centimeter $= 3.937 \times 10^{-9}$ inch $= 0.0001$ micron.

array—an assembly of solar cells usually mounted side by side on a flat surface and interconnected for high output.

B

beam—A narrow pencil of light rays (*see* ray). When the rays come together at a point beyond the light source, the beam is convergent. When the rays spread apart along the length of the beam, the beam is divergent. When the rays are parallel throughout the length of the beam, the beam is coherent.

blackbody—An ideal body or surface visualized as having zero reflection; that is, all light striking such a body would be absorbed. If an ideal blackbody could be achieved in practice (it can be approximated), it would be a perfect emitter when heated.

brightness—*See* intensity.

brilliance—*See* intensity.

C

cadmium selenide photocell—A photoconductive cell (which see) in which the light-sensitive material is cadmium selenide (CdSe).

cadmium sulfide photocell—A photoconductive cell (which see) in which the light-sensitive material is cadmium sulfide (CdS).

candela—Abbreviated cd. Also called *new candle*. Unit of light intensity in the International System of Units (SI). It is defined as (1) the luminous intensity of 1/600,000 square meter of a perfect radiator at the temperature of freezing platinum, (2) the luminous intensity equal to 1/60 of that of 1 square centimeter of a blackbody surface (*see* blackbody) at the solidification temperature of platinum.

candle—Abbreviated c. *See* international candle.

candlepower—Abbreviated cp. Luminous intensity expressed as equivalent to the light of so many candles (*see* International candle). Example: a 32-cp lamp. This unit is not the same as the footcandle, which is a unit of illumination (illuminance).

cell frame—Another name for array (which see).

coherent light—Light composed of parallel rays.

collimator—A device which converts light into parallel rays. In the simple spectroscope (which see), the collimator consists of a cylindrical tube having a narrow slit on one end, through which light is admitted, and a convex lens on the other end, through which parallel rays emerge.

color—The perceptible feature of visible light that is directly related to wavelength. Because of this feature, the eye experiences a unique sensation for each wavelength in the visible spectrum; thus, light at the high-frequency end of the spectrum "looks violet," and light at the low-frequency end of the spectrum "looks red." A complete description of a color must take into account three attributes: hue, saturation, and brilliance.

Hue depends upon wavelength and is what most speakers mean by the word "color." Thus, red, yellow, blue, and so on, are hues. White is a mixture of all hues. Saturation refers to the purity of a color, that is, the extent to which the color is undiluted with white. Brilliance is the strength of the color. *See* intensity.

color filter—A light filter (which see) selected to pass or reject, as desired, light of a particular color.

colorimeter—A color meter (which see) which permits the identification of hue and intensity, especially by comparison with a standard. Colorimeters are useful in many fields, including chemical analysis.

color meter—A photoelectric instrument for identifying and/or matching colors. One form consists of a light meter (which see) provided with red, green, and blue filters that can be switched successively between the photocell and a specimen under study which is illuminated with monochromatic light. When identical readings are obtained for two specimens with the filters successively in place, the colors of the specimens match in hue, saturation, and brilliance. *See also* color.

color temperature—The temperature at which the spectral distribution of an incandescent source comes closest to that of an ideal blackbody (which see) at that temperature.

compound lens—A single lens made by cementing two or more other lenses together. Also, a system of two or more lenses positioned coaxially.

converging lens—A lens, such as the double-convex type, which focuses rays to a single point beyond the lens. Compare diverging lens. *Also see* lens.

corpuscle—A tiny particle. One of the assumptions of the quantum theory (which see) is that electromagnetic waves, such as light, can behave as particles.

D

diffraction—The bending of a light ray around an obstruction. The angle of diffraction is greater for the long wavelengths, and a greater angle is most likely to occur when the obstacle is small compared to one wavelength.

diffraction grating—A device consisting of a transparent plate or film containing a great many equally spaced parallel lines or grooves (up to 40,000 per inch, but commonly 15,000 per inch). Light rays pass through the resulting slits and arrive at identical points outside, where they produce interference patterns. The wavelength of light can be measured in terms of the interference pattern and the slit size. Gratings are employed in some spectroscopes in place of a prism. A grating produces a spectrum by means of interference, whereas a prism does so by means of refraction.

diffusion—The multipoint reflection of light from a rough or unpolished surface, or the transmission of light through a translucent substance. Both processes produce a spreading of the light rays, and this imparts a softness to the light. Light is diffused also by a concave lens.

dispersion—The breaking up of light rays into their various color components, usually by means of a prism or a diffraction grating (both of which, see).

diverging lens—A lens, such as the double-concave type, which diverges rays through itself from an apparent focal point behind the lens. Compare converging lens. *See also* lens.

Doppler effect—An apparent change in the frequency of light when the distance between the source and the observer is rapidly increasing or decreasing. The frequency is higher when the distance is decreasing, and vice versa. One instance of this effect is the "red shift of galaxies," noted by astronomers: the frequency of light from stars moving away from the earth shifts toward the low (red) end of the spectrum, whereas the frequency of light from stars moving toward the earth shifts toward the high (violet) end. A similar effect is noted with sound waves: the pitch of an automobile horn is higher when the car approaches then when the car retreats.

E

electric eye—A popular term for photocell (which see).

electroluminescence—Luminescence (which see) arising from the application of voltage to certain materials.

electroluminescent cell—A light source utilizing the phenomenon of luminescence (which see). In a typical cell, a sensitive phosphor (which see) is coated onto a metal backplate which serves as one electrode. A thin, transparent layer of metal is evaporated on top of the phosphor, to serve as the other electrode. Finally, a protective glass or plastic plate covers this metal layer. An applied ac voltage (115 V to several thousand volts, depending on the size of cell, kind of phosphor, and desired brightness) causes the phosphor to glow, the generated light passing through the transparent metal layer and the glass cover. When a dc voltage is employed, the cell emits only a single flash of light at the instant that the voltage is applied.

electromagnetic wave—Energy radiated through space or matter in the form of a wave motion having perpendicular electric and magnetic components.

emergent ray—The ray which leaves a lens, prism, or other transmitter after being operated on by the device.

exposure meter—A light meter (which see) designed especially for use by photographers and photofinishers.

F

filter—*See* light filter.

fluorescence—Luminescence (which see) that lasts only as long as the excitation remains.

focal length—The distance from the center or cross section of a lens to the principal focus (which see).

focus—The point at which rays refracted by a lens or reflected by a mirror converge. Also the point *from* which rays seem to diverge.

footcandle—Abbreviated fc. Unit of illuminance: 1 fc = 1 lumen per square foot and is the illumination present on a surface of 1 square foot every portion of which is 1 foot from a point source of 1 international candle intensity, the surface being perpendicular to the rays.

footcandle meter—A light meter (which see) in which the scale of the output meter reads directly in footcandles.

footlambert—Abbreviated fL. Unit of average brightness; 1 fL = 1 lumen per square foot. This unit has been largely replaced by candela per square foot (*see* candela) and lumens per steradian square foot.

frame—Another name for array (which see).

G

grating—*See* diffraction grating.

grid—In a photocell, any network of thin wires, plated metal lines, or fine screen on the face of a semiconductor layer with which it makes contact. The wires are spaced for the admission of light to the surface.

H

hue—The attribute of light that is commonly meant by the word "color." *See* discussion of hue under color.

I

illuminance—The state or extent of illumination, *i.e.*, of being lighted. The unit of illuminance is the footcandle (which see).

illumination—The state or condition of being lighted. Also, the light itself.

image—The visual reproduction of an object, which is formed by a lens, mirror, or prism. Compare object.

incandescence—The generation of light by an intensely hot body.

incident light—Light that arrives at, and strikes, a surface or target.

incoherent light—Light made up principally of nonparallel rays.

index of refraction—The constant $n = \sin i/\sin r$, where i is the angle of incidence, and r is the angle of refraction. It is equal to the ratio of the speed of light in one of the two media involved in the refraction to the speed of light in the other medium.

infrared radiation—Electromagnetic radiation at wavelengths just below the visible spectrum. The infrared region extends from 7000 to 10,000,000 angstroms, approximately. Compare ultraviolet radiation.

intensity—The strength or amplitude of light. Also called brightness, brilliance, and luminosity. Units of intensity are the candela and the international candle (both of which, see).

interference—The interaction between light rays, which produces reinforcements and cancellations, depending upon phase, when the rays meet each other. This effect is similar to interference between sound waves or radio waves. In the case of light rays, cancellations produce dark areas, and reinforcements produce brightened areas; these effects create typical patterns—such as bars, concentric circles, moiré, and so on—just as cancellations between sound waves produce quiet spots, and cancellations between radio waves create zero-voltage spots.

internal impedance—The terminal impedance of an unloaded photocell. Like other electric generators, the cell delivers maximum power output when the load impedance equals this internal impedance.

international candle—Unit of light intensity. The unit 1 international candle is defined as (1) the light emitted by the flame of a ⅞-inch-diameter sperm candle that burns at the rate of 7.776 grams per hour, or (2) the luminous intensity of 5 square millimeters of platinum at the solidification temperature.

isolator—Another name for optoelectronic coupler (which see).

J

junction photocell—A photocell (which see) containing a pn junction in a semiconductor body. An example is the silicon photocell (which see).

K

Kerr cell—An electrically controlled light modulator or light valve one form of which passes the light through an organic dielectric liquid such as nitrobenzene. When a voltage is applied, the material becomes double refracting and restricts the transmission of light through itself.

L

lambert—Abbreviated L. The cgs (centimeter, gram, second) unit of brightness: 1 L is the brightness of a perfectly diffusing surface radiating or reflecting 1 lumen per square centimeter.

laser—A device for producing intense coherent light through the optical pulsing of atoms in a suitable material. The heart of this device may be a suitable gas, such as helium, or a rod of solid material, such as natural or synthetic ruby. In the ruby laser, the rod is silvered at each of its optically flat ends, the back end being completely silvered so that no light can pass through the mirror it forms, but the front end being only partially silvered. The ruby atoms are stimulated by means of strong light from a flash tube. This light forces the electrons in the atoms to a higher energy level. When the flash ends, however, the electrons fall back to their original level and in so doing emit energy in the form of visible light. A small amount of this light escapes as random incoherent rays, but that which moves down the rod excites other atoms along the way, cumulatively increasing the amount of generated light. This light is reflected by the opaque mirror back down the rod to the transparent mirror which reflects it back to the opaque mirror. The two mirrors continue to pass the light back and forth between the ends of the rod, more atoms being stimulated and the intensity of the generated light increasing with each trip. Finally, a highly intense coherent (single wavelength and phase) beam of light passes through the transparent mirror and out through the front end of the rod.

lead sulfide photocell—A photoconductive cell (which see) in which the light-sensitive material is lead sulfide (PbS).

lens—A polished (usually glass, quartz, or plastic) disc or plate, one or both faces of which are properly curved to produce a desired refraction of transmitted light. Convex lenses produce converging rays; concave lenses produce diverging rays.

light-activated switch—*See* photorelay and photoswitch.

light filter—Any device that transmits a certain desired frequency (wavelength) of light while holding back all others or, conversely, that holds back a certain frequency while transmitting all others.

lighting—Another name for illumination (which see). Also, the equipment which provides illumination.

light meter—A (usually simple) photoelectric instrument used to measure illumination, commonly in footcandles. In its simplest form, the light meter consists of a photovoltaic cell (which see) connected to a dc microammeter.

light quantum—Another name for photon (which see).

light sensor—A generic term for light-to-electricity transducers, such as photocells, photodiodes, phototransistors, and light-activated SCRs.

light source—Any body, device, or substance that generates light. Light arises from all other bodies by reflection. Principal sources of light in nature are the stars (such as our sun) and other hot bodies (planets are only reflectors). The following are various common processes for generating light.

Incandescence is the generation of light by an extremely hot body. An example is the filament in an electric lamp.

Luminescence is the generation of light by any means other than incandescence and which accordingly can occur at "cold" temperatures. Luminescence results from the absorption of energy and is specifically named in terms of the kind of energy causing it: electroluminescence (neon lamp and other simple gaseous tubes, electroluminescent cells, light-emitting diodes); photoluminescence (laser); bioluminescence (light-generating insects and plants); and chemiluminescence (light-generating chemical reactions). Luminescence results when radiation —such as electrons, X-rays, ultraviolet light, or radioactive particles —strikes a sensitized screen or certain crystals (oscilloscope tube, tv picture tube, fluoroscope, scintillating crystal, fluorescent lamp). When light is generated only while the excitation is present, the luminescence is called fluorescence; when light persists after the excitation is removed, the luminescence is called phosphorescence. Cathode-ray tubes are basically fluorescent devices, but some exhibit phosphorescence, as well.

light valve—A device, such as a window of liquid crystal or a Kerr cell (both of which, see), in which light transmission is controlled by means of an applied voltage.

liquid crystal—A material, resembling transparent or translucent plastic, whose light transmission is altered by means of an applied voltage. The nematic type of liquid crystal is normally transparent but is made temporarily turbid by the voltage.

lumen—Abbreviated lm. Unit of luminous flux: a uniform point source of 1 international candle emits 1 lumen in a unit solid angle.

luminance—In a given direction, the luminous intensity of a surface, stated in candelas per square meter.

luminescence—The generation of light by any process other than incandescence (generation by intensely hot bodies). *See* detailed description of luminescence under light source.

luminescent cell—*See* electroluminescent cell.

luminosity—*See* intensity.

luminosity curve—A plot showing spectral response (which see).

luminous energy—Radiant energy transmitted as visible light.

lux—Abbreviated lx. Unit of illumination equal to 1 lumen per square meter, or the illumination of a surface that is uniformly 1 meter away from a point source of 1 international candle intensity.

M

meter-candle—Another name for lux (which see).

milliphot—A small unit of illumination equal to 0.92903 footcandle = 0.92903 lumen per square foot = 10 lumens per square meter = 10 lux = 10 meter-candles = 0.001 phot.

mirror—A highly polished plate or disc (or the prepared portion of some other device) employed to reflect light. The surface of a mirror may be flat or curved. Familiar curved types include concave, convex, and parabolic. The parabolic type is common in searchlights and reflecting telescopes. Many mirrors are made by silvering glass, but others are fashioned from metal, mica, plastics, and other materials. A transparent (or "two-way") mirror is obtained by applying an extremely thin coat of a suitable metal to a transparent plate; this coat reflects some of the light that reaches it and transmits the rest.

N

normal—An imaginary line perpendicular to the face of a mirror or prism. The angle of incidence and the angle of reflection are reckoned from this line.

O

object—The body which is viewed with a lens, mirror, or prism. Compare image.

opacity—A substance is opaque when it transmits no light; an opaque material thus may be regarded as a light insulator. A great many materials—such as wood, metals (except in thin films), stone, some ceramics, cardboard, and some plastics—are opaque in proper thicknesses. (Many naturally opaque materials become transparent or translucent if the light is extremely intense or the sample is extremely thin, or both.)

optical coupler—*See* optoelectronic coupler.

optical system—The combination of lenses, mirrors, or prisms (or a single one of these) and their accessory devices, used with photocells in optoelectronic setups.

optics—The science of light and its applications and of the instruments used in measuring light phenomena. Also, another name for optical system (which see).

optoelectronic coupler—A signal-coupling device consisting essentially of a light source (such as an incandescent lamp, neon lamp, or light-emitting diode) enclosed in a light-tight housing together with a photovoltaic cell, photoconductive cell, or other photoelectric device. An input signal activates or modulates the light source so that the light output is proportional to the input-signal current or voltage or is a replica of the modulation. The light rays then activate the photocell so that the output signal likewise is proportional to the input signal or reproduces the input-signal waveform. Under some conditions, the optoelectronic coupler provides signal amplification. This device gives almost perfect input/output isolation because of the extremely low capacitance between the light source and the photocell.

optoisolator—Another name for optoelectronic coupler (which see).

P

panel—Another name for array (which see).

phosphor—A material that emits light when it is excited by electricity or some form of radiation. Common examples are the substances used on cathode-ray tube screens and X-ray fluoroscope screens, on the inside of fluorescent-lamp tubes, and in electroluminescent cells.

phosphorescence—Luminescence (which see) that persists after the excitation is removed. Compare fluorescence.

phot—Unit of illumination equal to 1 lumen per square centimeter, or the illumination of a surface that is uniformly 1 centimeter away from a point source of 1 international candle intensity.

photocell—A device (usually two-terminal, flat-plate) for converting light energy into electrical energy or for controlling the flow of an electric current by means of light. There are two general types: photoconductive cell and photovoltaic cell (both of which, see).

photoconductive cell—A photocell (which see) with a resistance that decreases when the cell is illuminated, and which accordingly can be employed as a light-controlled variable resistor for the control of an electric current.

photoconductivity—The property in which some substances—such as selenium, cadmium sulfide, and lead sulfide—undergo a temporary change of conductance (and accordingly of resistance) under illumination.

photocurrent—An electric current resulting from, or controlled by, the action of a photoelectric device such as a photocell.

photodetector—Another name for optoelectronic coupler (which see). Also, a device, such as a light meter, for measuring light intensity in terms of the output current of a photocell.

photoelectric cell—The term once considered standard for any simple device (usually two-terminal, flat-plate) whose electrical characteristics are temporarily altered by illumination. The term has been largely supplanted by photocell (which see).

photoelectric effect—*See* photoemission.

photoelectron—An electron that is ejected from an atom by light energy.

photoemission—The ejection of electrons from any atom by action of light.

photometer—(1) A photoelectric instrument for measuring illumination, having essentially the same internal arrangement as a light meter (which see) but usually more accurately calibrated and provided with several ranges. (2) An optical instrument for measuring illumination, in which light from a source under study is visually compared with that from a standard source, such as a calibrated lamp.

photon—A small packet—or quantum—of energy in which the electromagnetic energy that constitutes light is thought to travel. Photons account for the particle, or corpuscular, nature of light.

photorelay—A relay that is actuated by means of a light signal. The simplest form consists of a sensitive dc relay operated directly by a photovoltaic cell (which see). An electronic form has a photocell operating a triac, silicon controlled rectifier, or switching transistor.

photoresistive cell—Another name for photoconductive cell (which see).

photoresistivity—The inverse of photoconductivity (which see).

photosensitive—Responsive to radiant energy in the visible spectrum.

photoswitch—(1) An on-off device consisting essentially of a low-resistance, high-current photoconductive cell (which see) which is turned on by illumination and off by interruption of the illumination. (2) A photorelay (which see) designed to switch lights on and off.

photovoltaic cell—A photocell (which see) that generates a voltage when illuminated. Also called a self-generating cell.

Planck's constant—Symbolized by h. The constant 6.63×10^{-27} erg-second employed in quantum-theory expressions and important in many advanced studies of light phenomena.

point source—For purposes of analysis and design, a light source (which see) that is visualized as having no area; in practice, one that is so small—such as a pinhole—as to approach an ideal point source. Because of their great distance from the earth, stars appear as points of light to the unaided eye.

polarization—Ordinarily, light waves vibrate in all directions in a plane perpendicular to the line of travel of a ray. When the vibrations are limited to one of these directions, however, the wave is said to be plane polarized, or simply polarized. Thus, light may be horizontally polarized or vertically polarized, to mention two instances. Some polarization results from reflection or refraction, but most often light is polarized by transmitting it through certain crystals (such as calcite, quartz, tourmaline, mica, a Nicol prism, or a Polaroid filter). Some substances rotate the plane of polarized light passed through them. For example, a solution of the sugar dextrose rotates the plane to the right, whereas a solution of the sugar levulose rotates the plane to the left.

principal axis—The imaginary line joining the centers of the two spheres of which the two surfaces of a lens are sections. The principal focus (which see) lies on the principal axis.

principal focus—The point on the principal axis (which see) at which rays emerging from a converging lens meet, or at which rays behind a diverging lens seem to meet. Each lens has two equidistant principal foci, one in front of and one behind the lens.

prism—A block of glass or transparent plastic, having three flat faces giving the block a triangular cross section. Light undergoes refraction as it passes through the prism, and it is broken up into its various color components as it emerges from the prism. *See* dispersion, refraction, and spectrum.

Q

quantum theory—The theory that holds, among other things, that electromagnetic waves can act like particles. Thus, such particles, or "quanta," called photons in radiated light collide with and displace the electrons in photoelectric materials. Some phenomena—such as diffraction, interference, and polarization—are best explained when light is regarded as waves. Other phenomena—especially the photoelectric effect and fluorescence—are best explained when light is regarded as consisting of particles.

R

ray—A thin line of light, visualized as passing from the light source or reflector and continuing in a straight line until obstructed.

real image—An image (which see) that appears to the viewer to be on the other side (*i.e.*, behind) the lens or mirror. Compare virtual image.

reflection—Some of the light that strikes a surface is neither absorbed nor transmitted, but is bounced back. This phenomenon is termed reflection. An efficient reflector, such as the shiny, smooth surface of a mirror, returns virtually all of the incident light. In reflection, the angle between a reflected light ray and the normal (an imaginary line perpendicular to the surface of the reflector) equals the angle between the normal and the incoming (incident) light ray; that is, the angle of incidence equals the angle of reflection.

refraction—The bending of a light ray as it passes obliquely across the junction between two substances, such as air and glass, or air and water. This bending results from the difference in the speed of light in the different substances. In a familiar example of the phenomenon, refraction causes a stick that is partially immersed in water to appear bent; it also causes a submerged object to appear closer to an observer on shore than it actually is. Different wavelengths of light are refracted in different amounts by a substance. Because of this, light is broken up into its various color constituents when it is refracted by a transparent prism.

In refraction, the angle between a refracted ray and the normal (an imaginary line perpendicular to the surface of the refractor) equals the angle between the normal and the incoming (incident) light ray; that is, the angle of incidence equals the angle of refraction. An important characteristic of all materials that transmit light is the index of refraction, n. This quantity is expressed as:

$$n = v_1/v_2$$

where,

v_1 is the velocity of light in the first medium,
v_2 is the velocity of light in the second medium.

The index of refraction generally is expressed with respect to air or vacuum. Some common values of n are: air, 1.00029; glass, 1.5 to 1.7; and water, 1.33. Common refractors are lenses and prisms. Convex lenses produce converging rays; concave lenses produce diverging rays. *See* convergent and divergent under beam.

S

saturation—The attribute of color that expresses the degree of white that is mixed with a color. See discussion of saturation under color.

scintillating crystal—A plate, disc, or film of crystalline material, such as sodium iodide or phosphorescent zinc sulfide, which emits flashes of light when exposed to radioactive energy. Such a disc is the heart of the scintillation counter, a sensitive radioactivity-detecting instrument in which a photosensitive device (usually a photomultiplier tube) is activated by the flashes and delivers equivalent electrical pulses to an amplifier/counter circuit.

selenium photocell—A photocell (which see) in which the light-sensitive material is a thin layer of processed selenium deposited onto an aluminum or iron disc or plate. This photocell is usable as either a photoconductive cell or a photovoltaic cell (both of which, see).

self-generating photocell—Another name for photovoltaic cell (which see).

shingle-type photocell—A multiple unit in which several separate photocells are connected in series by slightly overlapping them in the manner of roof shingles. This arrangement gives the advantages of a larger cell without introducing the higher internal capacitance of the larger-area unit.

silicon photocell—A junction-type photocell in which processed silicon is the light-sensitive material. In this device, an n-type silicon layer is applied to a metal backplate which becomes one output electrode. A thin p-type layer then is formed on, or diffused into, the exposed face of the n-type layer. Finally, a metal collector ring (or a transparent metal layer) is applied to the p-type layer and serves as the other output electrode. The silicon photocell is primarily a photovoltaic device.

solar battery—(1) A group of solar cells (which see) connected in series, parallel, or both for high output. (2) A single, large photovoltaic cell (which see) that delivers substantial output voltage and current.

solar cell—A photovoltaic cell (which see) that provides notably high output voltage when exposed to sunlight.

solar panel—Another name for array (which see).

solar power—Electric power—usually assumed to be in useful amounts—obtained directly from solar cells rather than from sun-heated steam boiler/generator systems.

spectral response—The response of a photocell or the human eye to light of various wavelengths (colors).

spectrograph—A spectroscope (which see) in which the human observer is replaced with a camera for obtaining permanent records of observations.

spectrometer—A complete spectroscope (which see) with provisions for making quantitative measurements (such as those of wavelength) on observed spectra.

spectrophotometer—A photometer (which see) that can be adjusted to various wavelengths in the light spectrum for separate measurements of individual components. This instrument is analogous to the continuously variable tuned rf or af circuits used in electronics.

spectroscope—An optical instrument for producing and studying the spectrum of (usually) visible light. It employs a collimator (which see) which produces parallel rays from the source of light and which transmits these rays to a prism or a diffraction grating (both of which, see) which disperses the light into a spectrum (which see). The spectrum and a wavelength scale are viewed simultaneously through a self-contained telescope. A common use of the spectroscope is the analysis of chemical elements or compounds from the spectra of the light resulting from their burning. The composition of stars is determined in this way.

spectrum—Plural, spectra. (1) A continuous band of frequencies or wavelengths. (2) The visible spectrum showing the seven component colors of visible light and their intermediaries.

speed of light—*See* velocity of light.

sun battery—Another name for solar battery (which see).

T

total reflection—The complete return of a ray by a reflecting surface. This ray arrives at the surface at a critical angle of incidence, and virtually none of its energy is either absorbed or transmitted by the reflecting medium.

tracking—The continuous (usually automatic) positioning of a cell or array so that it follows the apparent movement of the sun across the sky.

translucence—A substance is translucent when it transmits light only partially. Paper and frosted glass are common examples of translucent materials. A translucent material may be regarded as a light attenuator.

transmission—Light that passes through a material is said to be transmitted by the material. Much of the light that strikes a transparent material is transmitted, whereas only a part of that striking a translucent material is transmitted, and none of that striking an opaque material is transmitted. Transmission is the opposite of absorption. *See* absorption, opacity, translucence, and transparency.

transparency—A substance is transparent when it readily transmits light. Clear glass is a familiar example. A transparent material may be regarded as a light conductor, the opposite of an opaque material, which is a light insulator.

U

ultraviolet radiation—Electromagnetic radiation at wavelengths just above the visible spectrum. The ultraviolet region extends from 100 to 4000 angstroms, approximately. Compare infrared radiation.

V

velocity of light—In a vacuum, the velocity of light in round numbers is 300,000 kilometers per second, or 186,000 miles per second. It is only slightly slower in air, but the velocity is lower in all other media.

virtual image—An image (which see) which appears to the viewer to be in front of (*i.e.*, on the viewer's side) of the lens or mirror. Compare real image.

visible spectrum—The band of wavelengths in which electromagnetic radiation produces the sensation of sight. This band extends roughly from 4100 angstroms (violet light) to 6500 angstroms (red light).

W

wave—Another name for electromagnetic wave (which see).

appendix **B**

Component Manufacturers

Photocells and Solar Cells:

B3M-C International

CL5M4L Clairex

CL505L Clairex

CL909L Clairex

CS120-C International

S1M-C International

S4M-C International

S7M-C International

S2900E5M International

S2900E7M International

S2900E9.5M International

SP2A40B International

SP2B48B International

SP2C80B International

SP2D96B International

SP4C40B International

SP4D48B International

Transistors, ICs, Diodes, SCRs, Varactors:

CA3010 RCA

CA3020 RCA

SK3009 RCA

U183 Siliconix

1N34A RCA

1N4815A TRW

2N190 General Electric

2N2712 General Electric

2N2608 Texas

2N3823 Texas

40810 RCA

Relays:

D1-960 Calectro

4F-5000S Sigma

5F-1000S Sigma

W588ACPX-16 Magnecraft

Miscellaneous:

AR-109 Transformer Argonne

AR-121 Transformer Argonne

AR-174 Transformer Argonne

EP50-C Motor International

P8626 Transformer Stancor

103 Counter Simpson

Identification of Manufacturers:

ARGONNE—Lafayette Radio Electronics, Syosset, L.I., NY 11791

CALECTRO—GC Electronics, Rockford, IL 61101

CLAIREX—Clairex Electronics, Inc., Mt Vernon, NY 10550

GENERAL ELECTRIC—General Electric Co., Semiconductor Products Dept., Syracuse, NY 10001

INTERNATIONAL—International Rectifier Corp., El Segundo, CA 90245

MAGNECRAFT—Magnecraft Electric Co., Chicago, IL 60630

RCA—RCA Solid State Div., Somerville, NJ 08876

SIGMA—Sigma Instruments, Inc., Braintree, MA 02185

SILICONIX—Siliconix, Inc., Santa Clara, CA 95054

SIMPSON—Simpson Electric Co., Elgin, IL 60120

STANCOR—Stancor/Essex International, Chicago, IL 60618

TEXAS—Texas Instruments, Inc., Dallas, TX 75222

TRW—TRW Semiconductors, Lawndale, CA 90260

Index